DURHAM CITY-COUNTY L
DURHAM, NORTH CA

OF
MOSQUITOES
MOTHS
AND MICE

BY THE SAME AUTHOR

A NATURALIST IN TRINIDAD

MOSQUITO SAFARI
A Naturalist in Southern Africa

OF
MOSQUITOES
MOTHS
AND MICE

C. BROOKE WORTH

DISCARDED BY
DURHAM COUNTY LIBRARY

W · W · NORTON & COMPANY · INC ·
NEW YORK

DURHAM CITY-COUNTY LIBRARY
DURHAM, NORTH CAROLINA

221425

Copyright © 1972 by C. Brooke Worth. All rights reserved.
Published simultaneously in Canada by George J. McLeod
Limited, Toronto. Portions of this book first appeared
in *Audubon*. Printed in the United States of America.

FIRST EDITION

Library of Congress Cataloging in Publication Data

Worth, Charles Brooke, 1908–
  Of mosquitoes, moths, and mice.

    1. Insects—New Jersey.   2. Zoology—New
  Jersey.
I. Title.
QL475.N5W66   591.9'749   75–39017
ISBN 0–393–06390–9

1 2 3 4 5 6 7 8 9 0

CONTENTS

Chapter 1

MOSQUITO ALLURE

When a sea captain steps ashore for the last time, he traditionally establishes himself in a home facing the breaking waves, for the chances are that he is not yet estranged from maritime attitudes but is simply no longer fit to act on them. Perhaps a valid parallel to that familiar lingering allegiance can be drawn in my own case, though I must confess at once that it concerns a love for mosquitoes. That assertion will very likely draw immediate and total ridicule without a shred of charity reserved toward the notion that such insects may possibly be worthy of a lifetime's veneration.

Yet it is so. At least that was part of my scheme in selecting southern New Jersey as a site for further enjoyment of nature after surviving intensive stints of mosquito study for the Rockefeller Foundation in India, South Africa, and Trinidad in the West Indies during parts of two decades. Surely the "Jersey Mosquito" is as famous as any in the world of tropical medicine. What could be more apt, then, than for an ex-malariologist and ex-arbovirologist to feel compensated in his recent abdication of the globe by the comforting presence of a faithfully buzzing company of voracious gnats amid scenes of his youth?

I had not grown up in New Jersey, but that is only a technicality, because I spent almost all my summers on the shore of Cape May County and could bear personal witness to annual increments of real estate blight affecting sand dunes and salt marshes for half a century. But that's something for later attack.

Now I want to testify in behalf of my long-standing acquaintance with Jersey mosquitoes. I used to squander my days on the vast salt marshes between Avalon and the Shore Road, during earlier years merely looking at Clapper Rails and fiddler crabs but later adding some purpose to enjoyment by banding nestling Ospreys for the old Biological Survey (now the Fish and Wildlife Service). One endured a multiplicity of vicissitudes under the hot sun, of which mosquitoes were only the beginning. Green-headed flies bit like stilettoes, while chiggers entered one's clothing and injected irritant spit in the tenderest places. The decorative linear cedar islands where Ospreys nested were labyrinths of catbrier, honey-suckle, and poison ivy, with holly and a ground cactus thrown in for prickly good measure. Even in the open marshes one's sneak-ers might be cut by mussels that grew vertically up through the mud in concealed fat clusters.

That was before the days of insect repellants (except for dubiously effective citronella), and I realize, now that I come to think of it, that things were worse for me because I was wearing short pants. Those two apparently unrelated observations come together by virtue of the practice I employed to protect my legs from both chiggers and winged biters. First I would wrap each spindly shank in toilet paper and then, as I gradually worked my long stocking up over the barber-pole spiral, I dusted in copious spoonfuls of flowers of sulfur. The laborious stratagem did only a partial job, mostly because the toilet paper either slipped down or else disintegrated if it became wet. Besides, that accounted only for one's person below the knee. I wore long-sleeved shirts, but it was not possible to line such a garment with paper armor. When I was climbing a cedar tree to inspect a nest, all my ap-pendages were in use simultaneously and I could not swat nor brush biters away. Then I might afford free drinks to a hundred mosquitoes on my back and a couple of dozen on each arm.

Therefore when I bought a farm at Eldora in Cape May County, New Jersey, it was not by any means in ignorance of the South Jersey mosquito. Yet there was something very peculiar about the circumstances surrounding the transaction. As time went on, I felt that the atmosphere was almost ominous. I had been taken by a real estate agent from Cape May Court House to the last farm in the county on the Delaware Bay side, on what we call Delsea Drive. On the way there he had said, "I hope you like mosquitoes." I told him that of course I did, and that that was one of my reasons for wanting a farm in this area.

It did not take me long to come to terms with Rudd Kolarich —his farm was exactly what I wanted. "I hope you like mosquitoes," he said, even before we had agreed to the sale. I repeated my assertion of positively adoring mosquitoes, and Rudd and the agent both smiled wisely. This was in late September in a year of drought, by the way. A few mosquitoes were biting us even as we stood and talked about them. Had the season been wetter or the date somewhat earlier, I would have expected more of the pests. But I found the two gentlemen's separate references to mosquitoes—and especially their manner of referring to them—somewhat mystifying.

Rudd became and remains a great friend as well as my chief aide on the farm. He began his role as assistant immediately by taking me around the neighborhood to meet everyone of importance. Mrs. Veach, the postmistress in the local office serving Eldora, said, "I sure hope you like mosquitoes, Doctor." So did Jacob Siegel at the food market in Woodbine, and so also did Herman Rosenfeld at South Jersey Farmers Supply in the same community. Now I knew that there was something that I did *not* know. But with the onset of cold weather, I was automatically denied finding anything out until the following year.

There was plenty else to find out about anyhow. Sixty-three acres is a lot of ground if you want to become intimately acquainted with every square foot of it. Indeed, that cannot be done and you have to settle for a very superficial awareness that some of the land is there. In the case of my salt meadows, the distant or aloof relationship could be accepted as sensible, for in venturing to explore them one was sure to sink into mud over one's shoes at every step and possibly up to one's hips in certain concealed holes. As for the woods, I am confident that there are still some acres I have not trodden, simply because of their number *and sameness*. The expression "If you've seen one, you've seen 'em all" applies with particular aptness to the relatively sterile woodland areas of South Jersey. Those not actually in the acknowledged Pine Barrens region lie adjacent to it and show themselves less barren to only a modest degree. Not that those monotonous forest galleries are without charm—I could in fact extoll them with what you might consider exaggerated feeling.

Salt meadows and woods: that was where I would hunt for game from algae to orchids and from amoebas to whatever spectrum of vertebrates awaited discovery. For the rest, my land included a large cultivated field (which one of my neighbors utilized in an extraordinary agricultural cycle that culminated in frozen bush lima beans) and another ten acres or so on which sat the house and some outbuildings amid a kitchen garden so vast that Rudd used to pay his annual taxes on the place by selling excess vegetables to passing motorists. My first act was to order demolition of the vegetable stand, but I nevertheless continued to plant the garden on an only slightly diminished scale; it embarrasses me to think how many friendships I have strained by dumping great loads of turnips and parsnips on people who hate turnips and parsnips. (I have found, by the way, that almost everybody does.) But they are so foolishly easy to raise—and delicious, to me at least—that I can't resist putting them in each year.

In the sea-captainly sense, then, I had really cast a connoisseur's eye on the salt meadows as well as the woods when I bought the place, rather than regarding the marshland as only so much incidental wastage that happened to go along with the rest of the property. My experience with tropical mosquitoes had universally established them as breeding in almost every conceivable habitat. Therefore the more varied the environment and the types of water collections to be found in it, the more elaborate would the mosquito fauna prove to be. In temperate areas there are fewer kinds of mosquito to begin with. It was all the more important, therefore, to select a non-uniform situation in order to increase the chances for variety. The famous Jersey mosquito went also under the name of salt-marsh mosquito, a term of significance to me now, for this and all other arguments I could think of urged the importance of my acquiring tidal meadows, while the possibility of encountering mosquitoes from freshwater sources as well could still be entertained. The main reservation in my mind, after all those leering remarks about my "liking mos-

quitoes," was that perhaps I would ultimately find that my meadows were providing a bit more fun than I had asked of them, as
if the sea captain one morning found his shore cottage afloat on a
tidal wave.

Strange to say, I had never looked carefully at a Jersey mosquito. I had not examined one under a magnifying lens to learn
the distribution of dark and light scales on its head, palpi, proboscis, thorax, wings, legs, and abdomen, nor did I know the relative
proportions of those anatomical members. It was either shocking
or ludicrous, depending on how seriously you take South Jersey,
that I could glibly recognize all the common mosquitoes of southern Africa, and most of the rare ones, too, but did not know a
single New Jersey species at sight. Well, this was one of the things
I had come "ashore" to do, and on many cold evenings of that
first winter I pored over a large volume called *Mosquitoes of
North America* by Stanley J. Carpenter and Walter J. LaCasse.
Those authors give descriptions and ranges of every species
known north of Mexico, in addition to full-page illustrations. I
was thereby able to abstract a list of the forms to be expected in
New Jersey and also to gain some idea of their appearance and
habits.

As for the "salt-marsh terror," they had the following to say,
although in this case my stock of prejudice needed but little re-
inforcing.

"The adults are strong fliers and often migrate in large numbers to communities many miles from the salt-water marshes in
which they breed. . . . An occasional female specimen has been
captured in light traps at distances as great as a hundred miles
from the salt-water pools in which the larvae are known to develop. The females are persistent biters and will attack any time
during the day or night. The adults rest in the vegetation during
the daytime and will attack anyone invading their haunts, even
in full sunlight."

But, of course, one needed further criteria for identifying
this creature with certainty. Even a fiercely aggressive daytime-
biting swarm of mosquitoes might occasionally be composed of
more than one species or of a totally different species—after all,
you cannot identify any kind of mosquito by the *feel* of its bite

alone. How, then, could *Aedes sollicitans* (for that is the strangely
spelled name of my current protagonist) be distinguished and
separated within a mixed feeding swarm by standard entomological methods? From Carpenter and LaCasse's diagnostic keys I
was able very readily to derive a few simple guides, each of
which could easily be followed to the appropriate anatomical
target through a 5× magnifying glass. Of course one knew in advance that these were not anopheline mosquitoes because the
palpi flanking a proboscis were short and stubby instead of equal
in length to the piercing organ. And one knew also that in the
non-anopheline sector these were *Aedes* mosquitoes, rather than
belonging to some other genus such as *Culex* or *Mansonia,* because the tips of their abdomens were pointed instead of rounded.

Now one had to get down to distinctions among sister species
of *Aedes.* Yet the task was still so facile as better to be called a
pastime. From *Aedes mitchelli,* for example, *A. sollicitans* differs
in having a yellow median ring on the first segment of the hind
tarsus. How could one pinpoint such a minute item? Very easily!
Under the lens, the leg stood out clearly as having a proximal
segment, the femur, or quite logically a "thigh"; then a tibia, or
"leg"; and finally a five-segmented tarsus, or "foot." In human
anatomy we homologize the tarsus with our ankle, but in Greek
"tarsus" means the "flat of the foot," so the term is more appropriately used by entomologists. Anyhow, having instantaneously disposed of femur and tibia of the hind leg, one comes directly to the
first segment of the tarsus and spots a yellow ring at its midpoint:
Scratch out *A. mitchelli!*

The key thus not only serves to eliminate other species. At
the same time it gradually builds up a description of the creature
you are specializing in. Accordingly, I learned that *A. sollicitans*
is characterized not only by the yellow ring around the first segment of its hind tarsi but also by wings with intermixed dark
brown and white scales; abdomen with a white to yellow dorsal
median longitudinal stripe (this turned out to be an extremely
handy mark for rapid identification); proboscis also with a ring
near the middle, though this was white rather than yellow; and
some additional light markings on the tarsal segments of all legs.
What the key, and the extensive description in the text, did *not*

tell, but a characteristic which I found striking, is that A. *sollicitans* appears to have a slightly upcurved proboscis, like an upturned nose. This gave the entire bearing of the insect a frivolous, or perhaps a delinquent, aspect, and it was quite readily understood that a convoy of them on a sortie could be up to no good.

As a matter of fact I simplified my identification problems, after selecting only those species known to occur in New Jersey, by condensing the keys in the book to cover that sparser fauna. That maneuver scaled my study down to forty-three or only about one-third of the total North American list. These were distributed among the following: seven *Anopheles*, one *Wyeomyia*, one *Uranotaenia*, three *Culiseta*, two *Orthopodomyia*, one *Mansonia*, four *Psorophora* (also having sharply pointed abdomens but differentiated by other means); three *Culex;* and the rest and by far the greatest number—twenty-one *Aedes*.

That still sounded like a formidable roster. In truth, it *was*. New Jersey is, despite what Texans may think, a fairly large state. Or perhaps I should emphasize that it has a conspicuously diversified terrain. The northwestern corner isn't so terribly far from Scranton, Pennsylvania, which is getting up into the coal regions, and that's a far cry from pine barrens or salt marshes. If the state supported forty-three species of mosquito, the chances were that few if any of them occurred throughout the area because of wide variations in climate and physical features. These irregularities would affect adjustments of winged adult mosquitoes as well as the available types of natural terrestrial collections of water for development of larvae, resulting in a scattered type of distribution of the species. Eldora, as one small spot in Cape May County, and my farm as an even smaller subdivision, must certainly support only a portion of the state's fauna, perhaps even a negligible fraction. Possibly A. *sollicitans* so dominated the local airways that other species gladly relinquished saline meadowlands to it and I had only a single form to recognize.

I was sure I knew what the first spring mosquito would be. Well, I *thought* I knew, and my alleged knowledge went back to a very concrete and reliable experience. During World War II I was stationed at the Army Medical School in Washington, D. C., and one early spring I was ordered to accompany one of my parasitologist colleagues, Captain Reinard Harkema, on an expedition to collect larvae of *Aedes canadensis* in Raleigh, North Carolina, for teaching purposes in our medical entomology laboratory. The reason for going to Raleigh was that "Hark" came from there, and he would consequently know where to look for larvae. We failed, and this is how I remember my facts: It was said that we were too late, since *A. canadensis* is the very first woodland pool species to develop each year.

During that first spring on the farm I exposed myself to mosquito attack by sitting on logs alongside teeming pools as soon as the weather became moderately warm. However, it was not until April 26 that I was able to watch a flying needle as it settled on the back of my hand, drove in its pointed stylets and began to engorge. Naturally I carried with me an aspirator tube especially made for capturing minute insects. It was equipped with a fine-mesh screen to prevent my aspirating the mosquito all the way down into my lungs. I ran back to the house with my catch, blew it out into a jar with a few drops of carbon tetrachloride to kill it, mounted it on a tiny steel pin and then viewed it raptly through my new binocular dissecting microscope with zoom focusing device. Immediately I saw the pointed abdomen—*Aedes*, yes—but the lack of a dorsal longitudinal stripe on the abdomen meant that it was not *A. sollicitans*. Good! I was now truly in the woodland pool business, and my carefully prepared abridged key would receive its first test as I attempted to follow the specimen through to its correct identity.

There are two things about entomological keys that must be unconditionally accepted by those who use them: They are extraordinarily accurate and they work almost infallibly. They do not present mere approximations of the truth. Just because insects are tiny does not mean that their characteristics are more inconstant than those of larger organisms. We would be unwilling to endorse the statement that most Robins' breasts are red, although a few Robins have blue ones. A bird with a blue breast obviously belongs to some other species.

Consequently, I found myself mystified almost as soon as I had worked through the first couplet of the key. I came directly up against the choice between this mosquito's having tarsal segments white-banded only at their bases or white-banded at tips as well as at their bases. If it were *A. canadensis* (I cheated by looking ahead!), the specimen should have both kinds of banding, but this one did not. As I have said, you must respect these dichotomies as you revere gospel. If my mosquito did not show up with the proper markings, it was not some peculiar or aberrant example of *A. canadensis*—it simply was not *A. canadensis* but something else.

Of course this was not really a tragedy. I didn't care what kinds of mosquito lived in my woods, so long as I could identify them and learn something about their life histories. That I had miscalled this one in advance did not make any particular difference except for its humbling demonstration of my unfamiliarity with South Jersey facts of nature. The key drew the specimen finally to a couplet which would make it *Aedes vexans* or *Aedes cantator*, depending on whether the dorsal surface of the seventh abdominal segment were dark- or light-scaled. This being a bright, freshly emerged female, I had no difficulty in seeing that her scales were pale. Thus, I was now able to welcome *A. cantator*, the "singing" *Aedes*, to my roster of woodland residents. Carpenter and LaCasse informed me that this is usually the dominant species in New Jersey marshes in the spring, but less common than *A. sollicitans* during the summer months.

Although *A. cantator* thus qualifies as an early spring form, it does not appear as *the very earliest*. According to the book, *A. canadensis* remained a quite reasonable possibility to have expected on my farm. The puzzle to me, then, was whether *A. canadensis* might have been present but had escaped my observation. Could such a phenomenon be true? Let me skip ahead now to the following spring in order to give an answer to that one question. This time I decided not to wait until I was bitten but actually to raise mosquitoes in the house. Soon after snow melted and forest pools took up their old positions, I ladled out some of the tannin-tinted water, as well as several rotting leaves and a bit of the bottom muck, and transferred the entire collection to a wide and shallow enamel pan on my dining room table. Immediately I found that I had brought in a world of diverse creatures—mosquito larvae to be sure, but also seed shrimps, copepods, water mites, isopods, and other sorts of water wraiths that gave me hours of entertainment as I watched them through a very handy dual magnifying glass, worn with a head band and called an Optivisor.

However, my chief interest lay in the mosquito larvae. These managed to grow, then to form pupae, and ultimately to emerge as adult insects of both genders. Since only female mosquitoes bite (the males living on sugary plant exudations), one usually encounters only that sex and their spouses are generally never seen unless one looks especially for them, when even then they may be hard to find. Mosquito eggs, on the other hand, yield the "normal" ratio of male and female individuals, namely equal numbers of each. Therefore if one rears larvae, one is assured of an ample supply of males. As a rule these members of the family, rather more fragile than their sisters and possessing handsome bushy antennae, are no more than curiosities to the practical entomologist. But it so happens that there are a few instances in which males must be examined to determine the identities of certain mosquitoes. Here the larvae as well as females look alike and the only way the two kinds can be sorted out is by examination of the male genitalia. The tip of the abdomen of a male mosquito ends in a pair of hooked claspers which he clamps onto the female's abdomen during copulation. These claspers, as well as the actual external sexual organ between them, bear numerous hairs, scales, tufts, and spines in addition to thickened chitinous processes that have served entomologists exceedingly well as landmarks for bestowing taxonomic names. The function of these structures is mysterious, that is, one can understand clearly how any male mosquito uses them, but there is no apparent reason why the elaborate details of various species should differ so conspicuously from one another. Here again one must accept the entomologists' assurance that the differences are constant: If the key states that a certain kind of male has such and such a spine situated just so on his claspers, that is how all normal males of that species will look. And yes, Carpenter and LaCasse had keys to male terminalia as well in their inexhaustible book.

I was not about to add this frill to my study, because it re-

quired a complex mounting technique and equipment that I did
not want to have cluttering up my lab. But I was glad to add
pinned males to the collection; this somehow rounded out the
more general type of information that anyone might like to know
about a given species of mosquito, and the females also looked a
little less remote and unfriendly with a few males flanking them.
Thus I soon acquired a quite respectable series of whatever it
was that struggled nakedly through metamorphosis in my dining
room as I watched practically every act with my nose all but in
the water.

"Whatever it was!" What should it be but my missing *A.
canadensis?* Thus the most obvious advantage of all in rearing
mosquitoes from their larvae is that you get the mosquitoes! Apparently *A. canadensis* had been emerging from my woodland
pools—probably even last year, too—but they had then gone
about their business oblivious to the attractions of mankind as a
source of blood, or actually disdaining those attractions in favor
of some other host. I have previously pointed out how clearly *A.
canadensis* proclaims itself with its doubly banded tarsal segments.
I could not possibly have missed it, had ever so worn or damaged
a specimen come my way among a crowd of *A. cantator.* No, the
species had simply *avoided me.*

A passage in Carpenter and LaCasse confirmed the tenor of
my observation without, however, advancing an explanation for
it. "Ross describes the females as being persistent biters [in Illi-
nois], attacking readily in shaded situations throughout most of
the day. Mail made similar observations on the biting habits of
this mosquito in Montana. The senior author [Carpenter] has observed that *it is seldom a troublesome biter in the eastern part of
its range* [italics mine], even in areas where large numbers have
recently emerged." The latter item was challenging enough to
launch any ardent entomologist on a life's work, though he might
even at its beginning feel pessimistic about arriving at a full
answer to all the questions raised. He might ferret out the preferred hosts of *A. canadensis* throughout its range and compose a
chart on which the relative palatability of Man as a mosquito-
sugarplum could be expressed in terms of the percentages—
compared with other hosts—attacked in Illinois or Montana as op-

posed to Ohio, Pennsylvania, and so forth. But he could never find out *why*. If you moved all those people from Illinois to South Jersey, they would not suddenly induce South Jersey *A. canadensis* mosquitoes to change their dietary habits; they would not taste sweeter than lifelong Jersey Crackers. The human beings would be basically the same, wherever they came from. On the other hand, I venture to guess that if Illinois *A. canadensis* were brought to South Jersey, they *would* bite mankind to the same extent as they did in the Middle West. Thus it is the mosquitoes that must be basically different and—once more—the "why," the nature of the difference—these are questions so fundamental but so terribly difficult as to be next to insoluble.

I had a further very pleasant experience with *A. canadensis*. Indeed it was also a ludicrous episode. However, I did not mind, because I was really not trying to impress anyone to much of an extent but was mainly having a private good time. What I did, in short, was to make a discovery that had already been made. When you do this in science, you can always cover up by calling it "confirmation" of someone else's results. But in the usual case, the other person's work is already known to you and your effort to confirm it was both intentional and deliberate (unless perchance you set out to prove the opposite but failed). In this case, I freely admit, I was not at all aware that I was duplicating an already recorded observation. Consequently I experienced the joy of chalking up a new fact, and I honestly did not feel much cheated later when by sheer chance I happened to read about "my" discovery under another author's name in the *Journal of Medical Entomology*.

One afternoon near the middle of June I was making the rounds of my forest bird nets. My route took me along the edge of the grain field (pertaining to frozen bush lima beans). Suddenly I confronted an eastern box turtle, and since I am inordinately fond of turtles, I immediately stooped to greet it. To my great titillation I saw that it was being harassed by a retinue of about twenty mosquitoes. Box turtles usually stop when they meet a human being—indeed they very often hiss and withdraw themselves entirely into their hinged shells. If this one had done so, I imagine that its attendant mosquitoes would have continued buzzing about it or even settled down to await its re-emergence.

This turtle, however, kept plodding right along, for reasons that I can not name. I doubt that it was because the mosquitoes were annoying it, because a turtle certainly can't outrun any biting pest. More likely it was merely because turtles don't always stop, and this was one of those turtles or one of those times. Its continuing progress made a more interesting tableau for me. As I have intimated, the mosquitoes had no trouble keeping up with it, but they were having a somewhat difficult time with their biting. If you were a mosquito, you would learn that a proboscis cannot be stuck into a turtle just anywhere at all. The shell is, of course, almost impregnable, though I have seen ticks that managed to get their mouthparts through seams in its horny layer where margins of two scutes came together. So perhaps a mosquito could find a vulnerable spot or two in the forbidding extent of a carapace. But the reptile's head also is quite effectively armored, while the front and hind legs and the tail are heavily scaled and, from the standpoint of the scales themselves, are likewise virtually impervious. The mosquitoes found it necessary, therefore, first to alight on the turtle's head or appendages and then to probe for soft skin between the horny shields and scales. The turtle's stumping gait made this difficult. Most of the swarming mosquitoes showed no signs of having succeeded in feeding even partially, while those with half-distended abdomens proved by their equally enthusiastic pursuit that they had been interrupted at dinner but were not yet nearly ready to abandon the table.

I believe it scarcely possible that this turtle's activity could be interpreted as a behavioral adaptation designed to interfere with or retard the rate of the mosquitoes' success in blood-letting. So far as I could tell, the turtle was oblivious to the insects, giving no heed to those that managed to pierce its tough hide and dislodging most of these only through accidents of its own movements or the jostling of encounters with green stems in the line of its journey.

By coincidence I had seen another box turtle while picking peas in the vegetable garden scarcely a half-hour previously. I

had greeted it at its own level, too, and would surely have noticed mosquitoes flying around it if they had been there—even
without my glasses. The grain field adjoined my woods, and *that*
turtle was actually on a fringe joining those two habitats, while
the one in the garden had reached essentially open country except for the artificial cover of vegetable rows. It is unlikely that
one box turtle should attract mosquitoes and another be eschewed
by them. I proposed to myself that the explanation for this
case must be related to the unwillingness of turtle-minded mosquitoes to venture more than a wingbeat or two away from sheltering groves near their natal pools.

However, I must return to the turtle in the garden immediately to check it over carefully. If only a few mosquitoes had
followed along with it, I might have missed them after all. On the
other hand, I must also return to the house at once to snatch up
an insect aspirator tube and fly back to the turtle in the grain
field before it wandered so far that I could not find it again. How
could I accomplish both those urgent ends at once? Quite logically all I had to do was run like hell, and sure enough that made
everything easy. The garden turtle *was* mosquito-less. The other
one had, in fact, not moved at all when I breathlessly reached it
again. This had allowed several mosquitoes to get half-drunk on
it, and I was pleased to collect two or three of them while they
had their proboscides inserted into turtle skin and their bellies
were distended with blood. I did not see how anyone could reject
that as evidence of turtle-feeding. I hereby aver that *I* had not
been bitten and that those mosquitoes had not simply alit, blood-
laden, on the turtle after they had finished with *me*. But perhaps
I am being more insistent than necessary on the honesty of that
little swarm.

Anyhow, it was now with great pleasure that I returned to
my primitive lab, mounted the new specimens and "discovered"
that they were *A. canadensis*. Those broad white bands embracing both sides of the tarsal joints looked mighty pretty to me. Of
course I would have been happy no matter what species they
were, merely because it would be so intensely rewarding to know
what kinds of mosquito bit turtles on my farm. But that it should

be *A. canadensis* immediately "fit" the fact that the species was present, although I had not been bitten by any of its members. The problem already suggested by my reading of Carpenter and LaCasse could now be made all the more interesting by interchanging not only mosquitoes and people between Illinois and South Jersey but also a few shipments of turtles.

But wasn't there something wrong with the diagnosis? How could this "earliest" mosquito still be on hand to annoy turtles in my woods in mid-June? Apparently they were not breaking any rules. I read further that a few *A. canadensis* regularly remain present until late summer. It seems that there is only a single, more or less simultaneous hatch during the first warm days of the year, but some of those mosquitoes are able to survive for several months. During that time the females are busy seeking blood meals and ripening successive batches of eggs, which they deposit on leaf litter at moist edges of woodland pools. If these new eggs become inundated by shortly ensuing spring or summer rains, they do not respond by hatching—otherwise there would be continuously emerging generations of *A. canadensis* right up until frost, and only very few eggs would remain to carry the destiny of their species through the coming winter. Instead the eggs of the new year go through most of their embryonic development and then lapse into a suspended biological state, called diapause, in which metabolism practically ceases. They can now not be brought back to life, as it were, until they have undergone a deep-freeze experience. Thus spring and summer eggs simply accumulate on the forest floor, each in its turn achieving the proper internal condition for entering diapause, so that when winter arrives the supply of progenitors for next year's crop is spread prodigally underfoot as one crunches through the snow.

No—nothing was amiss. The date was acceptable, the identification beyond any possible question. I "sat" on my fine discovery smugly and wondered, from time to time, how soon I would share it with other entomologists by writing a short note for one of their journals. I would choose the simplest of titles: "*Aedes canadensis* Feeding on the Eastern Box Turtle." My ruminations extended into an imaginary opening paragraph, in which I set myself up as a rather shy and distinctly modest observer who would surely be embarrassed by the acclaim that must result from what I was about to disclose.

But then I suffered the shattered toy. Or perhaps it was more like the ruination of a solitude in which one's isolated thoughts are unassailable law. At any rate, I made an unwonted contact with the outside world by reading a scientific journal and ran immediately into an account of *A. canadensis* feeding on Blanding's turtle. Not only that, the author cited two prior publications that recorded engorgement of this mosquito on painted turtles and box turtles as well. The latter two notices appeared in 1965, ten years after Carpenter and LaCasse's book, thus escaping the more general publicity they would have received if they had predated the hardcover deadline.

Anyhow, I was now so far down the "Me, too" line that it was not worth telling anyone about my discovery. Nevertheless I retained the warmest feeling of gratitude toward my turtle and its private parade of pests. Many sudden insights, even though commonplace, are so personal that they carry an inherent freshness to the individual who is at their center. My first glimpse of the Taj Mahal was likewise as if no one before me had ever sensed and then advertised the beauty of that structure.

The disenchanting article, by the way, cited turtle feeding in both Wisconsin *and Illinois*. Thus there must be a meeting ground of human- and turtle-biters in the Midwest, and perhaps my proposed translocation experiments should involve the exchange of mosquitoes between South Jersey and a really far-western population such as those living in Colorado or even the state of Washington.

I now had my spring farm mosquitoes—*A. canadensis* and *A. cantator*—fairly well taped (though, come to think of it, I did not yet know where *A. cantator* came from). What else would the forest pools bring forth? I ought really to have built large mosquito-netting cages, at least six feet square and several feet high, to place over some of those sylvan water surfaces. Then everything that emerged would have been simultaneously trapped and I would be able to amass an authentic census of winged creatures originating in each thirty-six square feet of pool area. Perhaps I shall still do that. But at present my information is restricted to the much less exciting and thoroughly inefficient dining-room table technique. The enamel pan yielded plenty of *A. canadensis* —I had them flying all over the house—but as I peered into the microcosm through my Optivisor I thought sometimes that I saw larvae that looked slightly different from the average. These could not be moved while they were still larvae, for it was necessary for them to range about the pan and feed on whatever microscopic objects they required for their nutrition and growth. But as soon as they transformed to nonfeeding pupae, I ladled them out with a tablespoon and sequestered them in individual glass vials, with a bit of water to enable them to float and a square of fine-mesh netting tied over the top to prevent their escape.

One of those larvae didn't really look much different—its only distinction was a slightly larger size. Perhaps it was merely a particularly robust specimen of *A. canadensis*. But very much to the contrary! Its pupa somehow gave rise to a quite perceptibly larger mosquito. However, that was a minor point. The mosquito itself appeared under my binocular scope as a most beautifully caparisoned one. Its wings were thickly adorned with both light and dark scales in a salt-and-pepper pattern, each scale broad and paddle-shaped so that the overdressed insect achieved almost a mothlike aura. The legs and body showed up in a corresponding display of elegance that had me immediately revising my notions of drab aedine fashions in the Temperate Zone. You can easily imagine that such a mosquito almost ran itself through the identification key. *Aedes grossbecki* seemed a rather unimaginative name for it, inasmuch as I would have thought that the entomologist who christened it might have engaged in a bit of appreciative description. *Aedes mirabilis? Aedes pulcherrimus?* Something of that sort.

I might as well mention a few mosquito names at this juncture, because there *have* been several entomologists who took obvious pleasure in being humorous or facetious in their otherwise dull work of publishing detailed written descriptions of newly discovered species and providing them at the same time with titles that must last them for the rest of scientific eternity. The fact that one of those gentlemen spelled "solicit" with its classical double "l" is an incidental embellishment above and beyond the point he was trying to make. In addition to *A. sollicitans* and *A. cantator,* the genus *Aedes* contains (among a host of mere workaday names) *vexans, irritans,* and *excrucians,* while one discovers even greater intellectual efforts in the roster of *Culex* species, e.g., *vorax, horridus, inflictus, invidiosus, perfidiosus, abominator,* and *inadmirabilis* among many similar imaginative tours de force.

On the other hand, I am certain that Mr. (or Doctor or Professor) Grossbeck was inordinately flattered when he received his mosquito namesake. Unfortunately I can not report what he did to deserve it. Patronyms are awarded for many reasons—to pay homage to a scholar, teacher, or colleague; to acknowledge financial support or hospitality accorded the mosquito collector; and so on. Then there is *Anopheles barberi.* I happen to know that this native species (which is not a malaria carrier) was named for a Mr. Herbert S. Barber who, though an entomologist, was not a "mosquito man" at all. He encountered the first known specimen as it bit his arm on Plummer's Island in the Potomac River a few miles above Washington, and took the trouble to collect it because he happened to have a "mosquito friend" at the U. S. National Museum whom he wanted to please. Though Mr. Barber had no idea that the mosquito was anything but an ordinary one, his learned ally, Mr. Coquillett, immediately recognized it as a new species and duly immortalized Mr. Barber by exercising the entomologist's ultimate gesture of courtesy.

It is easier for me to think of reasons why mosquitoes should be plain than why they should evolve fancy markings. The same is true of all insects, actually, but it strikes me most forcibly in the case of mosquitoes because of their small size. The really attractive pattern of *A. grossbecki*'s broad, light-and-dark scales should, when one first thinks about it, have evolved primarily because members of that species took a delight in it, at least from a sexual standpoint, though whether that meant that males admired females or vice versa—or both—does not particularly matter. From a less romantic point of view, a bright color scheme might be beneficial merely by aiding rapid recognition, though in that event one would expect all species to have evolved distinctively until an assemblage of them looked like an international gathering of flags.

But all such introspection breaks down when one considers mosquitoes' eyesight and remembers that they can barely see one another *as mosquitoes,* let alone recognize a color design. The facets of a mosquito's compound eye are disposed on a roughly hemispherical surface in an arrangement that allows one insect, regarding another from a respectful distance of three inches or so, to pick up its neighbor's framework in possibly forty or fifty dissimilar fragments. None of these separate pieces is even a part of some larger image, for although each facet of an eye has a tiny transparent lens, this serves only to concentrate light rather than to focus it, there being in the depths of each optical unit no suitably curved sensitive surface to receive a clear picture of outward objects. Indeed, if each ommatidium, as the units are called, had to deal with a tiny but entire image rather than a mere point of light, the mosquito would suffer from *too keen* vision. An approaching enemy, such as a Purple Martin, for example, would be seen so many times and from so many slightly different angles that the nervous equipment needed for coordinating all the information would require a voluminous brain totally out of proportion to the creature's ability to house it and carry it about. Instead, a mosquito must make do with a very general idea of what lies in front of its eyes, seeing objects at best as a mosaic of dots of varying luminosity. That is why I marvel at the beauty of *A. grossbecki.* Such finery is *not* for the species' members to appreciate, for they could not possibly register the shape and hue of each other's scales. Then for whom did the pattern evolve, and why?

Carpenter and LaCasse do not answer that kind of question. I read, however, that *A. grossbecki* is "generally rare" in its range throughout the eastern United States. That statement gave me much satisfaction, just as bird watchers get a double thrill when uncommon birds are beautiful as well, like Blackburnian Warblers for instance. Not much is known about this mosquito, though its origin in woodland pools in early spring was already well established, while one observer had reported its feeding on human beings. Eventually I reared another specimen in the house, but thus far I have not been pursued by adults in my woods.

It was easy to be sidetracked by such novelties. But after all I was still faced with an unsolved mystery. When I got around to looking it up, I found that *A. cantator* is another type of "salt-marsh" mosquito, and consequently not to have been expected as a woodland product. Therefore I facilely presumed that I possessed an excellent breeding ground for both that species and *A. sollicitans* in my fifteen or so acres of tidal meadows. *Aedes cantator,* again with Carpenter and LaCasse's license, flew characteristically in early spring; its prompt appearance in my woods surely must denote an adjacent aquatic nursery.

And I might as well now pull out all the stops and report what happened a little later when *A. sollicitans* put on the show first hinted at by the real estate man, Rudd Kolarich, Mrs. Veach, Jacob Siegel, Herman Rosenfeld and the rest. Because when *that* number of mosquitoes was swarming about, *while I still could dip nothing but fish out of my meadows,* I realized that something peculiar must be going on.

Possibly it would be a mistake to pull out *all* the stops, for then no one would believe me. Perhaps I can make the point better by referring not to myself but to my wife, Merida, and to other people less intimately involved in the situation. That, by the way, is a good choice of words, for the other people managed not to be "intimately involved" by visiting me once *and once only,* while Merida simply scheduled her trips in winter. I have plenty of relatives and friends who like gardens, and my trips to my other home in Swarthmore with carloads of produce inspired them all to come down to the farm for a look at the luxuriant seed-catalogue tableau encrusting my hydroponic soil. But as each one walked from his car toward the first row of eggplant or zinnias, I could see him begin to stiffen while the clouds engulfed him, just as if he were entering a wall of flame or the noxious atmosphere of some alien planet on which his spaceship had landed a moment ago. In the next instant he was either back in his car or in the house. To observe the native reaction, watch Rudd as he rides my tractor-mower through the young orchard. It is a hot day in July or August, but he is clad as if for midwinter, wearing a visored cap with earflaps, a handerkerchief over his nose and mouth, heavy gloves, an impervious leather windbreaker, at least two pairs of pants, and knee-high rubber boots. Besides that, he is reeking of mosquito repellent and carries a leafy switch to dislodge the dozens of insects that manage to get past the various barriers he has set against them. I dress the same way as I go about my nature studies, and I can add the subjective note (which is, I guess, my last stop after all) that I find myself breathing very shallowly to prevent inhaled mosquitoes from getting more than a wing-length beyond the portals of my respiratory tract.

Well, this is vivid or purple reporting for a usually cautious mosquito buff. However, I shall soon bolster what has been said by quoting some illuminating entomological statistics. Not that statistics are always necessary to back up assertions or allegations. In this case one can easily skip over such figures and proceed directly to the For Sale signs and locally depressed real estate values. If people want to move away because of mosquitoes, it is not necessary to make a count of the pests before admitting that they are superabundant. (I'll mention real estate again after a few more excursions into wetlands.)

The point to which my unbelieving mind was driven did not make sense at the time: Salt-marsh mosquitoes were being produced in salt marshes roundabout but not in *my* salt marshes. Yet that was so, and great was the day when I learned the answer! From another standpoint I had been "unbelieving" in a hesitant way, for I did not know whether to trust memories of my younger days. Everything I have said about knowing the Jersey mosquito was based on experiences that now lay in a number of rather remote decades. Could several more recently interposed doses of tropical mosquito bites have distorted those older impressions? Did I really suffer then ten times as acutely as I now remembered? Again, the revelation that my farm was the site of ten—or ten times ten—as many mosquitoes as I had ever seen before in one place was bracing news, since it proved that I still had some ability to make accurate comparisons among my treasury of itches.

On page 97 of the *Proceedings* of the Fifty-third Annual Meeting of the New Jersey Mosquito Extermination Association,

which was held in Atlantic City, March 23–25, 1966, appeared
the following title: *A Summary of Nine Years of Applied
Mosquito-Wildlife Research on Cumberland County, N.J., Salt
Marshes,* by Fred Ferrigno and D. M. Jobbins. This has been
my "Chapman's *Homer,*" ever since I ran into it, especially because I could spit over my property line into Cumberland County
if I were willing to wade into the middle of West Creek. The
sixteen-page article has much to say about insecticides, but I
shall skip those parts for the moment. What commanded my
wonder at once, and what still remains as the explanation of my
"mosquito farm" (as others see it), is the description of an agricultural practice of which I had never heard and which is as
truly an example of modern-age specialization as is, for example,
root-canal dentistry.

It seems that entirely without my knowing anything about
it, many of my neighbors (in Cape May County as well as
across the border) were making their living by raising salt hay.
There are two kinds of salt hay—I knew that much—each with
several uses. The coastal variety is much coarser and is sometimes called straw, while on the Delaware Bay side (*my* side of
the peninsula) it grows, relatively speaking, more in the manner
of a dense, short, and rather fine grass. In the past both kinds of
salt hay were gathered where they grew wild or were raked up
in places where their broken stems had been piled up naturally
in windrows. The material thus formed a sort of bonus for farmers
whose property adjoined salt meadows. I suppose they began
gathering the crop for their own use at first, after someone discovered that when salt hay is used as a mulch—on strawberry
beds, for example—it serves better than ordinary hay or leaf
mulches because it contains no weed seeds. No doubt its own
seeds are there but cannot sprout or grow apart from a marshy
substrate. But in modern times the demand for salt hay has
extended to all sorts of unexpected industries: as an overlay for
curing new concrete on roads and similarly on floors of sky-
scrapers; as a packing for all kinds of shipping and freight; as
upholsterers' stuffing—and so on. I had often seen trucks, piled
almost to toppling height with bales of salt hay, passing in front
of the house, but I had given them no thought as *mosquito indi-*

cators. Why should anyone have made that crazy metal leap?

However, Ferrigno and Jobbins cited salt-hay culture as the major reason for *un*reasonable mosquito populations. They studied tens of thousands of acres of tidal meadowlands along Delaware Bay in Cumberland County and came up with more figures on pest production than even the most vicious statistician could have demanded. They found that in natural marshes, mosquito larvae were all but eliminated by minnows that they called "killifish," just as I had observed in my own meadows. Here and there in the marshes they found low-lying pockets, cut off from tidal flushing. In these, beyond the reach of killifish, mosquito production shot up to "between 1 and 6 million per acre per brood." But such pockets were scattered and did not necessarily embrace more than a fraction of an acre apiece. The "natural" marshes, therefore, accounted for mosquitoes more or less as I would have foretold them from past experience.

Present-day salt-hay farming calls for procedures that the old-timers knew nothing about. Not only is there no further dependence on beneficent windfalls, the farmer no longer trusts tides and the weather either. At a certain time early in the season he dikes in his hay meadows to create a semipermanent summer pond. The grass loves this and grows more lush than it ever did years ago. The process is reminiscent of rice culture, with its flooded paddies during young stages of the plants, except that the salt-hay farmer does not have to set out his crop painstakingly; verdure arises annually with seeming spontaneity from the underlying eternal sod. Also as in rice culture, the land must be drained before harvest. Now the farmer opens his dikes and lets the water run out. By haying time the ground has become firm and he can move in heavy equipment to cut and bale his large reward for having labored a little.

When dikes were first closed in the early part of the season, tidal action was promptly eliminated. That meant an end to killifish invasions, as well as death to those that were already in the impoundment, for the sun soon raised temperatures to the intolerable level of hot soup. But larvae of *A. sollicitans* found those conditions immensely stimulating. This must be just what the natural marsh pockets were always like. But instead of just an

isolated nursery here or there, Cumberland County now pos-
sesses ten thousand acres of diked salt meadows, the entire area
having been converted "into an enormous mosquito breeder" according to my newfound authorities. Let's see, now: at 6 million
mosquitoes per acre per brood, and assuming there are four
broods per year, I think that comes to 240 billion mosquitoes in
Cumberland County alone. And so it transpires that today people
are moving away and the wildlife is only a shade of what it used
to be. But there are many other angles to measure before we can
see the complete design.

Of course I was an old hand at the public health "angle,"
that is, the real, down-to-earth or down-to-the-grave dangers of
mosquitoes apart from their importance as nuisances. And although the public health aspects of the problem were almost lost
among other considerations in Cape May County, they definitely existed, and I shall speak of them first because of their
relationship to past voyages of the "retired sea captain." One of
my great concerns in the West Indies had been eastern equine
encephalitis (or EEE, for short). This mosquito-transmitted disease, which is not at all confined to equines, despite its name, has
been found most often in wild birds. Obviously when mosquitoes
and birds meet, within a virus problem, I try to squeeze into the
picture as a sympathetic investigator, just in case anyone tries to
say something nasty about birds for their part in the show. (I
can't always exonerate birds, but it is really up to the accusers to
make their point.)

If arthropod-borne viruses were equally readily transmitted
by all kinds of mosquito, we would be in a mess, if indeed we
were here at all. Fortunately, viruses are astonishingly choosey
about the hosts, both invertebrate and vertebrate, they will in-
fect. Just think how EEE virus would spread in South Jersey if a
quarter of a trillion *A. sollicitans* were waiting to give it a lift
every summer! In Trinidad, too, we dealt with large populations
of mosquitoes, some species of which were very abundant occasionally, but despite the presence of EEE virus on the island,
there had never been a known case of the disease in man and
only one in a horse, up to the year when I left (1965). There the
vector mosquitoes thus far detected had been *Culex nigripalpus*

and *C. taeniopus,* the former only periodically occurring in sudden swarms and the latter rare at all times. In South Jersey there was likewise reputed to be an uncommon transmitter, namely *Culiseta melanura,* and I use "uncommon" in both a relative and an absolute sense, in view of the fact that I have been searching for this species on my farm for several years now without finding a single one. But then, *A. sollicitans* is quite a discouraging haystack.

Nevertheless EEE is able somehow or other to manifest itself from time to time in South Jersey, either by being reintroduced from a more southern region or, perhaps, by maintaining itself locally in some mysterious way even in seasons and in years when no recognized cases of EEE infection occur in human beings or other vertebrate animals. For occasional cases *do* occur. Indeed, a man residing less than four miles from my farm died of eastern equine encephalitis one summer even as I went about my acres revelling in the delights of natural history. In fact I had recently put up a sign at my driveway entrance, reading NO SPRAY, to prevent poisoning of my property by the township-operated DDT vehicle, and the nearby tragedy might be thought a signal for removing the notice. But I felt that the spray truck was ineffective against mosquitoes, with all those salt-hay meadows roundabout, while in my immediate purview there were many harmless organisms whose company I preferred to perpetuate rather than destroy indiscriminately along with a paltry million or two *A. sollicitans.*

Now we come to the harshest impact of mosquitoes on society, and this has to do with money. New Jersey calls itself the Garden State, but a fat fraction of its economy is concerned with beach resorts. Whereas mosquitoes are nuisances to summer visitors, particularly to bathers in bikinis, such pests can be grumblingly tolerated. As long as the individual is willing to use repellents, and also when he observes that the community is trying to help by sending around the DDT fogging machine twice a day, he will put up with mosquitoes. Possibly the state loses a little revenue because a few especially sensitive people don't come to the seashore lest they be bitten, but in general the hotels and cottages are booked fairly solidly in July and August.

However, let no one whisper anything about encephalitis!

Following the death of my neighbor there were some shocking allegations in the newspapers. The burden of the innuendoes was that the diagnosis had been hushed up temporarily in order not to panic South Jersey's late-August visitors. Time enough to make it public after Labor Day, when everyone had gone home and sold-out concessionaires were closing their boardwalk stalls for the winter. But time, too, implied the newspapers, for some uninformed visitors to have become infected: Should not the diagnosis have been proclaimed in large headlines at once?

All this, unsavory though it was, had a basis in precedent. On past occasions EEE's advent was publicized and the resorts had lost ruinously. International quarantines may be economically disastrous, too, and efforts to suppress information when outbreaks of infectious disease occur are sometimes correspondingly devious and notorious.

But now I became exposed to a much wider preoccupation with mosquitoes by practically the entire business community and, consequently, by politicians and thus, eventually, by voters. Through my long-standing friend, Dr. Ernest A. Choate, a retired high school principal living in Cape May Point, I was introduced to a number of Cape May County citizens, official and otherwise, who had recently discovered that they must cooperate henceforth on mosquito problems for any number of hitherto unrecognized reasons. As a property owner and an interested member of the community, I joined these new friends in their efforts to regulate the environment wisely.

I could begin the explanation almost anywhere, since it turns out that everything is interrelated, but having already considered salt hay, let's take that for a starter. Salt-hay cultivaion produces an excessive number of mosquitoes, as we have seen, and mosquitoes do all sorts of negative things to the economic balance. But salt hay leads to other woes also. If mosquito larvae in the impounded meadows are to be killed, tremendous amounts of insecticide must be used to make up for the lack of natural larval predators. All this chemical burden makes the area unfit for other wildlife, provided the material stays put, but some of it drains into adjacent land and water tracts, contaminating those as well.

And quite apart from insecticides, even *untreated* salt-hay impoundments become unprofitable habitats for marsh birds such as rails and ducks, while the dikes effectively prevent access to black drum and other fish whose spawn require unimpeded acres of such nurseries. The hunter and the fisherman, to say nothing of the nature lover, therefore line up with the bikini-clad bather to condemn salt-hay meadows, for it seems that the only things these meadows are good for are salt hay and mosquitoes. Whereas the nature lover has a very feeble voice at the polls, hunters and fishermen can be heard clearly. Thus we have defined another point to which the authorities will listen with both ears.

In a democracy you can't abolish salt-hay culture by proclamation—or even by saying that you think it isn't nice. The farmers who now specialize in it have no other crop to fall back on; an arbitrary ban on the practice would simply proclaim them out of existence. The only solution seems to be to buy their land (which once would have been considered almost worthless) and restore it to its natural state by destroying the dikes. But today those acres will cost a lot of money, and the taxpayer is not yet nearly well enough educated to cast his vote for such a ridiculous appropriation. Who wants the county to acquire thousands of acres of expensive salt meadows while teachers' salaries remain inadequate?

However, there is another complexity. Though everyone looks narrowly at the salt-hay farmer, motives are not all alike. At least one group of people would like to have that "worthless" land, not to benefit the wilderness concept, but for its own brand of profit. Real estate developers have already been responsible for "reclaiming" vast acreages of coastal marshes, filling them in to make solid ground for building. For a long time nature was able to struggle along, thanks to the large supply of marshes and the slow pace of old-fashioned earth-moving operations. Today the situation is acute, tense. The real estate developer recognizes schizophrenic reactions in himself. He wants to sell to more hunters and fishermen who want more game. He must house them by "developing" the very marshes that are their reason for wanting to settle here.

The web extends almost indefinitely, entangling everyone

on the way. Antimosquito measures include drainage of all standing surface water, fresh as well as brackish or salt. Yet the State
Park Commission and the water resources experts both advocate
increasing freshwater storage in natural settings, not only for
recreational purposes but because the numbers of seasonal tour-
ists have already exploited available underground supplies almost
to the limit, and those incoming sportsmen (if they can be induced to remain) must find enough for their needs also.

It is clear that from now on no single group can have its un-
disputed way. The salt-hay farmer, the real estate promoter, the
pesticide salesman, the fisherman and hunter, everyone concerned with prosperity and comfortable living—all these people
must make concessions, even that greatest sacrifice of all which is
to allow a few mosquitoes to survive rather than sanctioning the
elimination of most marshes and poisoning of the remainder.

I feel that I may have made a rather hectic landfall after all,
for the mosquito seas remain turbulent in every direction. But
whatever happens—and I suppose there will be tidal waves before *A. sollicitans* is brought to bay—I just hope that lovely *A.
grossbecki* doesn't get caught in the undertow.

Chapter 2

CEDAR ISLAND

So engrossing are mosquitoes when they put their best efforts into battle that they occupy one's physical and mental attention alike. That explains why I have not yet mentioned strawberry flies. If you can imagine the experience of coping simultaneously with a pest that is just as bad as *Aedes sollicitans*, then you will achieve a perfect intellectual impression of life on my farm during the idyllic days of June, July, and part of August. I sometimes wonder why all the well-wishers did not add, "I hope you like strawberry flies, Doctor," when Rudd introduced me to the community. Actually, however, I think I can supply the answer. In some strange way you forget about strawberry flies over the winter. Mosquitoes come first in the spring, and then they last and last—long after frost, and don't let anyone deceive you about the benefits of a good October freeze: One year I squashed my last engorging *A. sollicitans* on December 9. Thus the memory of mosquitoes is persistent, while the torments of strawberry flies sink into an unremembered limbo only because these insects have the grace to quit after they have won instead of repeating their victory drive endlessly. I must commend them also for going to bed at night.

Just the same—and perhaps even more emphatically as a result of their unexpected return—the first flies of the season give one a nasty blow. At this time one is beginning to adjust to mosquitoes and is even wondering how it could be that last year they seemed so unendurable. Flies supply the answer: They push one far beyond that which can be tolerated. They bite and suck blood, of course, but they are exasperating besides, buzzing around one's eyes, ears, and nostrils, and crawling into the neck and sleeves of one's clothing. Obviously it is necessary to retreat, to get away anywhere at all, and just as predictably I drive ten miles across the peninsula to the fresh open seacoast where one can breathe great gulps of salt-scented relief.

At Merida's summer cottage in Avalon I can contemplate strawberry flies a bit more sanely. Now it is possible to make a few objective remarks about them. I am astonished to find that they are not mentioned by that name in any of my textbooks of medical entomology, for the term is understood by everybody in South Jersey and surely deserves to be noticed by compilers of pest lore. The insect—or insects, there being a succession of different kinds during the "fly season"—are easily recognizable as species of *Chrysops,* a genus embracing the deerfly branch of the horsefly family, for their wings are distinctly "pictured" with a dark broad crossband. Golden-eyed they are not, despite their Greek name; that description must apply to one or more of their sister species. Nevertheless the eyes *are* beautiful, while capable of sight reflecting iridescent metallic greens and reds amid black background mottlings, but losing all trace of glint soon after death.

Maggots of this family of flies—the Tabanidae—almost always develop in an aquatic medium, and I have no reason to suspect that those of strawberry flies depart from the usual practice. Moreover some species are adapted for life in brackish or even frankly salt water. Therefore, therefore . . . although I know nothing definite about the source of my own private farm scourge, some base instinct makes me look toward salt-hay meadows that lie just beyond an intervening stretch of woods to the south. Tainted already with one guilt, they could just as well be double-dyed. And again: I don't remember flies to this degree in my youth, though admittedly that experience was on the Avalon side; and today, as I relax in that haven, strawberry flies are scarcely noticeable.

Well, I came here to get rid of *thoughts* of torture as well as the physical ordeal of it. And, as always, my first impulse—or, more properly, the first gesture toward natural history—was directed at birds. A few doors away from Merida's cottage live the Liguoris, and on this occasion one of their sons, Victor, was there, eager to take me for a ride in his new boat. As usual I had brought bird bands with me, so after saying hello to my own kin, I found the offer a perfect opportunity to visit Cedar Island in a search for nestling herons of bandable age and size.

In Victor's boat this was quite an excursion, in spite of the modern advantage of an outboard motor. For we had to go down Princeton Harbor to the Inland Waterway, pass under the turntable bridge, and then continue north to the yacht club opposite which Cedar Island presented a convenient though very small cove with a man-made miniature sandy beach for dryshod landing. However, I wasn't worried about such niceties as clean feet. The sooner they got wet and muddy the better, so that I could forget about them. I'm not certain how Victor felt, though he followed me ashore quickly enough and dirtied his shoes without uttering a word. We had not slogged very far through the marsh, paralleling a row of cedars that grew on a low, straight ridge, when I saw several white birds floundering among the branches of one of the trees.

"They're either Snowy Egrets or Little Blue Herons," I shouted, halfway up the tree already. "They may be too old—not to band, I mean; but I may not be able to catch any."

I won't even pretend that you might have had the experience of chasing baby herons in their rookery, for that is one of the distinctly rare types of activity exhibited by our species and is usually engaged in by people who are too busy to read books. It is, I'll have to admit, a dubious pastime, particularly as the results of great effort are likely to be scanty or even nil. The young birds are able to climb almost as soon as they are born. With each succeeding day they become more adept at it. *Five* appendages come into play during such clambering—two legs, two wing tabs, and one skinny head and neck. Rookeries are invariably located in groves of trees or tall bushes, so that when one thinks one has just about reached a baby, it makes an incredible transfer to a neighboring branch, requiring one laboriously to descend and then climb the next tree, thereby often exciting the same dextrous exchange of perches by helpless-looking chicks. The babies take care to vomit and defecate on the climber—the number of direct hits is testimony of the intentional nature of those defensive acts. Incidentally, this gives one an opportunity to learn what types of food the parents have been bringing to their nestlings (fish and shrimps chiefly), though such information could be gained by more enjoyable methods and one need not argue that the babies have performed a gratuitous favor for science.

These white birds followed the typical pattern as described, but by persisting in my exertions I finally managed to grab one of them by the legs, just as it was about to flop across a chasm. I practically fell out of the tree with my prize and then was first able to get a good look at it.

"A Snowy," I told Victor. "They look almost exactly like Little Blues at this age, but the four outermost wing feathers of Snowies are pure white, while those of Little Blues are dark-tipped—not black, but dusky or dirty-looking. You can tell them apart by that one marking even before the feathers have sprouted from their sheaths, because the pinfeathers themselves have dull tips if the babies are Little Blues. . . . But here—hold the bird while I go back for my bands: I forgot and left them in the boat."

I thrust the egret, which was almost full grown and therefore about the size of a slender chicken, into Victor's grasp and ran through the marsh toward the little cove. It is easy enough to retrace your fresh steps in this kind of terrain, for the brittle glasswort and tough sea lavender do not spring back for a tide or two, while even the marsh grass tends to regard foot prints as insults to be forgiven reluctantly. I found it somewhat more difficult, however, to return in my mind to the first time I had left such tracks on Cedar Island and then to fill in all the intermediate visits. One thing was certain: This was the only time I had arrived by power craft. Of yore, I had exerted muscles on the oars of a rowboat or—at least once—in actual swimming, with my clothing, binoculars, and other gear held up in one hand out of the water.

That time—the swimming occasion—had been in 1928, when I was a student at Swarthmore College, and I am now amused to see that I purposely made my field notes as uninformative as possible so that no one could guess what had really happened. Usually when I go somewhere to look at birds, especially if it is for more than a one-day trip, I am careful to mention any companions who might have been along or else the people I met anywhere en route. But in this case there is nothing but a list of birds, the name "Avalon, New Jersey," and the indication that I arrived on June 8 and left on the eleventh. Heavens! That means I spent three nights in Avalon. Besides, early June is in the preseason era—at least it was in those days—so I could not have been visiting anybody. Surely this must have been an adventure. Then what was my reason for being cryptic about it?

The joke is now partially on me, because today I am unable to tell exactly who my co-criminals were—for we were indeed guilty of breaking and entering a summer cottage. It happened that Swarthmore and Haverford colleges each contained two keen young ornithologists in the same class, so that I was often afield with one or more members of the group. For some reason which I do not remember, we were rarely all on hand at the same time. Therefore while I am sure that two of them were with me in this Avalon venture, I cannot say which ones they were. Nevertheless it is mathematically certain that one of the boys was from Haverford College, and *that* is important because it was he who selected the cottage we broke into, saying that his parents were friends of the owners. On the other hand, I doubt that the families could have been such very close associates, considering how furtive he was about the whole operation and how we jumped every time a board creaked as if the cops were sneaking up on us. If I am correct, this scion later became a vice-president of the American Ornithologists' Union, but I'll not say anything else that might identify him, in case the statute of limitations has not yet run its course in releasing us from taxation for our sins.

The cottage could not have had a better location for our purposes, since it was built on the bank of a narrow channel whose opposite shore was Cedar Island's eastward expanse of salt meadow. The brown-shingled structure stands there yet, at 1918 Ocean Drive, though today it is much too large to be a "cottage" and has inevitably been cut up into apartments. It was called Avalon Lodge then. That was an admirable title, as we found when we made our way to the third floor. Here was a single huge dormitory-like room with a bank of ten windows facing east and a similar row of panes to the west. On one of the nights, when a howling thunderstorm buffeted Avalon, I felt that the cottage was about to be lifted from its foundations by winds from all four quarters.

As a matter of fact, I'm not sure there were any permanent police in Avalon during the off-season in those days. Anyhow, we were as secretive as possible in our comings and goings to and from Cedar Island, practically swimming under water to avoid being seen, and at night we lived in darkness, the electricity being shut off, of course, but also not trusting our flashlights lest they betray our illegal trespass. We used our unknowing hosts' beds (though without sheets) and we ate up all canned stuffs that they had been unwise enough to leave on hand. Not that we weren't prepared for camping or to pay our way fully in a more rustic style: It was simply logical to take advantage of bounties that came to hand so easily.

The purpose of this particular trip was to photograph birds at their nests, especially Clapper Rails, which are traditionally hard to observe at close range and therefore extra difficult to catch in camera poses. I had recently been notified of favorable action on my application to band birds for the Biological Survey, but the bands had not yet arrived. Therefore, since I had not yet taken up bird photography seriously, I was a member of this party in only two capacities, namely observer and assistant, positions which gave me the greatest possible latitude for *knowing* that I was enjoying myself.

Almost any of the salt-marsh islands along the South Jersey coast could be named "Cedar Island," since they share the two traits of being cut off from adjacent soggy or sandy land masses by mazes of channels that surround them completely, and of possessing a line or two of cedars growing on slightly elevated ridges—probably prehistoric sand dunes—that transect most of them in a north-and-south direction parallel to the outer seashore. Unfortunately many of these islands no longer exist or, if they do, their isolation has been nullified by a variety of causeways and bridges. Their cedars have become valuable as a source of driftwood, which all the summer visitors seem to covet. The trees grow in highly gnarled shapes to begin with, owing to their unprotected exposure to every storm, and I imagine that the efforts of various species of large birds to build nests in them do not aid in keeping erect postures. In any event, when older trees die by degrees, and broken pieces of them get picked up and carried long distances by high tides, they have truly become driftwood. But nowadays people are too impatient, or the demand is too great, and nature must be aided by those who invade such marsh groves to tear the trees apart prematurely. I suppose it is possible to etch the fresh pieces by artificial processes to make them look properly seaworn, weathered, or otherwise eroded.

In Avalon there has always been a particular island designated as Cedar Island, not that it was in any way superior, with larger or better trees or more birds to begin with, so I don't know why it should have claimed that undistinguishing name. Perhaps the most likely origin of the title was that this island lay a shade less accessible than others in Avalon, so that stress should be put on its insularity rather than its trees. Possibly for this reason, or very likely for other reasons of which I am not aware, Cedar Island escaped the indignities suffered successively by its sisters as one after another disappeared under cross-hatched streets and avenues. Thus the glimpse I can give of its 1928 aspect could pass for its modern image—and *that* is becoming an almost impossible thing to say for South Jersey landscapes everywhere.

Our approach from Avalon Lodge was to the southeast corner of the island. Where we struggled ashore over a mudbank pitted with fiddler crabs' burrows there was nothing but more mud, albeit level and grass-grown. The quarter of a mile between us and the first line of cedars now appeared to be a continuous sward, but that expanse was really filled with booby traps. Some of these were relatively honest ones, in the form of open pools that disclosed their margins quite generously, so that we eventually found our way around them all, even though they might have tortuous shapes. But the others! They were partially hidden side channels that ran all over the marsh, undoubtedly forming an effective system for leading in each tide and conducting it out again, so that virtually every square yard of the area received a periodic refreshing saline douche. But we did not know the details of all those interdigitating ditches, and to get from our point of landing to the cedar strip required us to run a naturally formed maze. Like so many laboratory rats we had to learn our way by trial and error. However, not being rats, we fumed over our mistakes and sometimes, saying the hell with it, we floundered across the channels in a ludicrously human way. I can aver that those barriers are much deeper and muddier than they look and it is greatly advisable to avoid them despite time lost in finding a land route around their devious and deceptive courses.

Cedars are the dominant trees from several points of view. Although none of them is of great height because their tops have a tendency to flatten out under atmospheric assaults, one can find plenty of individuals with trunks exceeding a foot in diameter at ground level, and these are probably the oldest residents of the strip. Younger candidates do not press upward in crowds, as one might suspect, but present themselves, still slender-tipped, unobtrusively among their seniors. The rate of replacement must be slow, for the most ancient cedars cling to existence sometimes for years through but a single viable branch, and aspirants for occupied root space would only waste their efforts by jostling.

Practically every Osprey's nest is borne by a cedar, usually at the flattened apex of a crown though occasionally on a large side branch. Conversely, almost every cedar of exceptional size and age carries an Osprey's nest, whether or not it is currently occupied. The relationship is scarcely symbiotic, for a nest obviously contributes to each tree's slow decline. But again the rate of changeover is slow, many trees remaining perfectly satisfactory nest holders for many years after their transmutation into inert standing driftwood, so that the gradual appearance of newly mature trees keeps pace with demands placed on them by the stately Fish Hawks.

Indeed the other significantly common trees of the ridge, also windswept and stunted, of course, are either too slight or else too impermanent to be of use to Ospreys, though they are often enough literally festooned with the rickety nesting platforms of herons and egrets. Hollies are sufficiently tough, to be sure, but they do not attain any measure of solidity on Cedar Island, that is, I have never seen one growing there that could bear up under a seven-foot wide nest weighing a few hundred pounds and capable of supporting me at its center. Wild cherries attain a greater girth than hollies but are both brittle and subject to rot. Only a foolish young Osprey would choose one of those trees as a foundation.

Around the fringe of this island and many others like it is a skirt of shrubbery formed by a continuous band of marsh elders. These bushes seem to be squeezed between the open marsh, which must be too wet for them, and the drier portions of the low ridges that are preempted by more formidable cedars, hollies, and wild cherries. Yet the marsh elders are nevertheless an important part of the structure of this interrelated animal and plant community. Their roots go down in an intermediate zone which is exactly where the higher tides deposit whatever they can move —salt straw in great masses of parallel leaves and shafts, true driftwood of all sizes and shapes, and since 1492 the buoyant castings that epitomize Western Civilization. Trunks of the marsh elders form a splendid anchorage for much of this material, and the entire botanical association of a cedar strip is thereby protected from erosion. Moreover the elders serve as cover for a contingent of smaller birds that inhabits surrounding marshes. Seaside Sparrows use them as singing perches, while Clapper Rails are likely to favor grass tussocks nearby or directly underneath the bushes, rather than a general marshy no-man's-land, for their nests.

That brings me back to photography and the sweating labors of my companions. Happily it was no problem at all to find rails or their abundantly dispersed, large clutches of eggs. Cedar Island supports actually two major low, tree-covered mounds, closely adjacent and parallel to each other, as well as a few smaller, broken outcroppings. Consequently the total perimeter of dry land, with its garnish of marsh elders, was of great length, possibly as much as a mile and a half or more, and a walk around that beat disclosed one nest after another. Therefore the photographers were able to choose among a number of different settings, selecting those that were situated most advantageously in relation to incident light at a preferred hour and also with regard to the positions incubating birds were likely to take as features of the projected composition of each picture.

After deciding on a particular nest for study, we had to put up a blind from which to observe it. This structure was a primitive shelter, perhaps no more than a piece of canvas propped up with a stick and covered with a few dead branches. Indeed the cruder it was, the more readily birds accepted it as a nonthreatening part of the environment. A peephole, alternately for an eye or a camera lens, did not upset the rails. Rather it was we who became upset after individually spending more than a few minutes in a blind, because the enclosure became stiflingly hot, while it was not permissible to exercise punitive reflexes against mosquitoes or green-headed tabanid flies that found us dandy picnic grounds as we lurked motionless inside. We evaded those discomforts by abandoning a blind to the camera alone, setting off the shutter by pulling a long string from a distance. Either telescopic lenses had not yet been invented, or else they were too expensive—certainly we had none. Nevertheless the cameras must have taken some excellent shots as we operated them—by remote control—in reasonable freedom from unbearable discomfort under the trees.

Siting of a blind in the correct position was invariably a cinch. Clapper Rails almost always approach their nests from the same direction and then squat over their eggs still facing forward, much as if they had entered the back door in preparation for bolting out the front. In doing so repeatedly, they wore trails in the grass that could easily be recognized as habitual runways. Hence if one wanted a side view of an incubating rail, it was necessary only to arrange the camera at right angles to the beaten track. For a head-on portrait one must, of course, determine which was the front door.

Then all that remained was to click the shutter. This, it seems, ought to have been the easiest chore of all, following the many other more complicated preparations. But now the rails themselves had to cooperate. In short, it was necessary for them to let us know where they were, and this sort of publicity was exactly what they intended to avoid. It is utterly amazing how a rail can practically run in circles around you at close range—in short cover, too—and still keep out of sight. This they do by "skulking," as all ornithologists who have written about rails are prompt to report. Skulking, I gather, is a particular type of slinking that includes the ability to squeeze through narrow passages between grass stems without causing the grass to move, a mistake which would give away the position of a bird as it progresses on its secretive way.

However, there was more method in the ruse than that. Such tactics might save many a rail's skin in the hunting season, when the skulking maneuver could be combined with a steady widening of distances between birds and intruders. Now the rails were motivated to remain and it was the invaders who would hopefully retreat. The best that could be done was to pretend that nests were nonexistent and that consequently all rails were living a life of carefree indolence. To get that point across, some birds deliberately exposed themselves along bare, muddy banks of marsh channels at some distance from the cedar strip. As I eventually learned, these each were likely to represent one half of a couple. While the visible member of a pair held one's attention by its conspicuousness, the other was busily skulking back toward its untended eggs. Often the return to a nest within only a few yards of one's feet was accomplished in complete absence of any external evidence of such homecoming.

Then, as I have suggested, why not simply click away? As a matter of fact, that is what we had to do. But the Clapper Rails were extremely tricky subjects. Sometimes they simply refused to return while we were about. Naturally we could not go to the nests to see if they were back, because that would set them off again and then there would have to be another wait of indeterminate length. Therefore when we clicked, it was with a greater measure of hope than any other emotion—certainly confidence was scarcely in it. I have said that lenses were expensive. So were films, and we were in no position to photograph large series of unoccupied nests for the sake of the rare successful shot. After pulling the string, we would have to go to the blind to reset the camera, and then if a bird scurried away it was possible to predict a positive image on the film. But often we could not tell a thing about the chances: An empty nest could mean that there had been no bird there; or else the timid rail might have returned but then been frightened by the noise of the camera and skulked away unseen before we arrived to make a new adjustment. Of course, if the eggs were cold, that would indicate that incubation had been suspended for a relatively long time—perhaps for hours —and this sort of interference was not included in our most enthusiastic quest for pictures. When it seemed that a pair of rails might desert their nest, we hastily deserted first, for there were plenty of other opportunities to search for less skittish parents.

I had had more striking encounters with these birds under a different sort of circumstance. When we were canvassing the island's margin for nests, but before a particular bird had been discovered, that incubating individual might take a chance on being overlooked and simply sit tight as we drew near. I presume that a few rails really did succeed in keeping the secret of both themselves and their nests in this manner. That maneuver would have been more effective, however, if we had not been specifically looking for rails but had simply bumbled along without purpose. As it was, we stuck to the zone of nesting concentration and thereby almost stepped on a majority of the cowering birds. These, now having waited for too long to be able to make invisible skulking exits, went to the opposite extreme, not only exposing themselves in a sudden wildly flapping flight, but squawking at the same time and, moreover, often coming down into the grass *close by,* where they continued in attempts to occupy our attention by various diversionary antics and calls.

Clapper Rails are inelegantly called mud hens by inelegant Crackers. However, the term is admittedly apt—if you don't look too closely. Almost any short-winged ground bird with big feet resembles a chicken to some extent. But even a nonornithologist would see at second glance that "hen," in the farmyard sense, has been applied to Clapper Rails under folklore license. This particular fowl has an entirely different sort of bill, relatively long, slender, and downcurved (the better to snap you up, my dear fiddler crab) instead of bluntly tapered, while the grayish brown plumage is much softer and more loosely worn than the harsh mail of ordinary poultry. Perhaps the marsh bird's voice has been invoked more strongly than its form in the establishment of chicken imagery, but here again one must not listen critically if it is desired to continue the fancy. A brooding Clapper Rail, with eggs near the hatching point, will sometimes move only a few feet from the nest, spread its wings as if sheltering chicks, and utter a rasping "kr-r-r-r-eck, kr-r-r-r-eck" that would never deceive a poultryman although it does have henlike overtones or undertones. However, one is most likely to hear a chorus of unseen birds at some distance in midswamp, and then it comes easily to accept them as a watery breed of barnyard fowl, for their sort of cacophony is surely not that of sparrows or larks.

Indeed the early ornithologists have written of the former abundance of Clapper Rails in the Atlantic coastal marshes in terms that suit the Mud Hen title better in their day than in ours. For this was then a "table" bird in both egg and carcass. The extent to which Clapper Rails were marketed, and at least a part of their decline *before* habitat destruction became widespread, must be inescapably correlated. Audubon writes of collectors who gathered 100 dozen eggs in a day, while he invites one to a mathematical computation that leads to an estimate of about 1 million rails in a salt marsh twenty miles long and one mile wide. As recently as the 1890s, Witmer Stone saw "gunners, back of Atlantic City, shoot them until their guns became too hot to hold and dead birds were left to rot in piles on the boat landings and adjacent mud flats." Such slaughter took place during high tides, when the birds could not skulk because their weedy shelters had become inundated. Another henlike characteristic that then appeared was their weak flight (though the species does succeed in migrating when at last it gets down to flying seriously). They must have been anything but challenging targets. In fact I saw two men catching Clapper Rails with large hand nets during a high tide at Avalon within the past year, though the birds were soon released wearing Fish and Wildlife Service leg bands. But that helped to make word pictures of hunts a century old suddenly become more vivid. One can easily place Clapper Rails in the category of creatures that are least prepared by nature to stand up to civilized man. That is particularly true of species adapted to specialized habitats. Salt marshes may appear vast at a rail's eye level, and Audubon's million birds per twenty square mile plot sound like hordes, but in actual fact the salt marshes are the thinnest possible strip along the continental margin, amounting virtually to a linear confine rather than vast two-dimensional expanses such as the eastern forests or central plains. Therefore *all* salt-marsh denizens must be regarded as endangered species, so long as dredges and draglines continue to huff and puff.

Fortunately the Clapper Rail is still a common species (by *today's* standards) so that the least of our problems on the Cedar Island expedition was nest finding. Actually we quickly became adept at identifying nests at a distance, even earlier ones that had already been vacated after the end of incubation. The birds often weave together a few grass tips over the nest to form an open-fretted canopy that can serve no purpose but an esthetic one, so far as I can interpret it. Certainly these gracefully arching wisps are insufficient to conceal either the sitting parents or the unattended eggs from hawk-eyed gulls passing overhead, while from ground level the structure, slight though it is, gives the secret away to one who knows what it means. Probably the practice has a meaning that has escaped me, but my present impression is that the birds may draw the grasses together out of boredom as they sit through the long days and nights. And, to be sure, there is no one to say that they don't think it makes home look like Minoru Yamasaki's celebrated, lacy, towering steel arches in Seattle.

When we disturbed a rail with eggs in the process of hatching, the old bird obviously determined to sit very close indeed and its final departure would be violently abrupt. After such a wild scramble, I sometimes found two or three jet-black chicks on the muddy ground about the nest, or perhaps actually hidden among grass stalks beneath the nest itself. It occurred to me after several such episodes that these stray chicks all had fluffy, dry down, while the nests still contained pipped eggs and *wet* chicks. The disorder, then, was not really so great as I had originally concluded, for shouldn't a clumsily frantic parent jettison nest items indiscriminately? Once again the old birds' performance was chiefly for our benefit—to draw our attention away from the nest —but it was nonetheless controlled. And the *dry* chicks were already practicing their first lesson in skulking. Since all the babies leave the nest as soon as the last-hatched one has fluffed out, the scattering of earlier ones during an emergency is not so remarkable after all.

I have not the faintest idea whether any of those photographs of rail domesticity are still in existence. I am not at all sure that I ever saw them back in 1928; if I did, they could not have made much of an impression. My excuse for such inattentiveness is that this excursion introduced me to a new ornithological venture which dominated my New Jersey summers for half a decade thereafter, and while my companions battled with cameras and rail behavior, I was already dreaming of future Osprey-banding exploits.

Cedar Island supported many pairs of Ospreys in those days, though unfortunately I have lost my record of the actual counts in successive years. There were always more nests than pairs of birds, for it might take many years for old abandoned eyries to disintegrate if they were not blown down first. My recollection is that Cedar Island itself might have been home for seven or eight breeding pairs each year, while the total census for this and all of Avalon's other similar islands might have come to twenty or so.

In any event, during a lapse in our furtive waiting, when it would not matter that we showed ourselves to the rails, I climbed an occupied cedar tree and peered over the rim of the nest at my first clutch of Osprey's eggs. The event impressed me from many aspects. Surely one of them was the *ease* which which one could reach such treasures. I have always been the world's worst athlete, while the tales I had read of people's struggles to reach hawks' nests had been cliff-hangers, both in actuality and in essence when trees, rather than cliffs, were involved. But this Osprey's nest was no more than ten or twelve feet from the ground, and the old cedar tree's dead lower branches, broken off to one-foot stubs, constituted a veritable staircase that would have allowed persons even feebler than myself to ascend with ease.

Another inviting facet of Osprey banding was based on the size of the birds. The records amassed by the Biological Survey were already voluminous enough to disclose some generalities

about the chances of hearing some day in the future about any
bird you might band today. Banded nestlings of small species
were the least likely to be recovered, presumably because they
would suffer a high mortality soon after fledging and were then
not apt to be found because of their smallness and rapid decom-
position. But as adult individuals of larger species were studied,
the rate of people finding their carcasses increased proportionately,
so that the great Fish Hawk, or Osprey, standing practically at
the upper limit of gargantuan dimensions, offered high rewards of
information from the most meager banding efforts. To put these
remarks into specific terms, one might expect only a small fraction
of 1 percent of the bodies of banded birds such as swallows or
wrens to be noticed and picked up by persons elsewhere, while
Osprey recovery reports might run as high as 10 percent or more.
Stated otherwise—and this really reveals the birdbander's chore
more strikingly—one would have to band hundreds upon hundreds of swallows or wrens before it was at all likely that the
mail would bring a coveted recovery report on one of them from
the office of the Biological Survey. But if one banded only one
nestful of Ospreys containing three young birds, the chances of
hearing ultimately about the travels of one of them were already
close to fifty-fifty; band two nestfuls, and you had a sure thing.

These considerations undoubtedly provided all the motivation I could possibly need to embark on a lifetime of Osprey
studies. Yet I was drawn—or pushed—by further stimuli, one of
which impinged, through my eyes, in that moment when I first
peeped over a nest's rim and beheld those huge eggs, blotched
with large splashes of deep chocolate brown. I have never been
a true "egger" (as bird lovers derisively call nest-robbing oolo-
gists), but that has not prevented my gawking into many un-
plundered nests simply to delight the eyes. Well, here was delight
aplenty. The nest itself, made of large sticks, was six feet or more
in diameter, with a broad, shallow concavity at its center that was
lined with dead, leafy branchlets, dried seaweed, and assorted
fragments of grass, reeds, denuded fish skeletons, cellophane, paper, rags, and other refuse. Somehow these formed a hospitable
receiver on which the eggs—only two of them in this case—reposed in harmony with their surroundings, their size first of all

looking "right" in a nest of such magnitude, and their coarse mottling complementing the background of trash so as to give the entire ensemble an appearance of natural good taste.

The birds themselves added a sense of tremendous adventure. When we had just invaded the island and were only beginning to flounder across the expanse of grass-grown mud toward the line of cedars, Ospreys rose from one nest after another to circle, screaming, over the marsh. Those whose nest happened to be nearest our advance became the most excited, hovering high over our heads and swooping intermittently over us, passing so close that we could hear the wind whistling through their wing feathers as they veered off at the end of a dive and seemingly "bounced" upward. Their voices were high-pitched and polysyllabic, so that the air was filled with a babble of complaining notes. To this was added a contrapuntal chorus from flocks of Laughing Gulls which, always on the alert for joining a disturbance, arrived almost immediately to circle with the Ospreys and have their say about us. (Herons nesting on the cedar strip were more circumspect. One by one, or in small groups, they simply abandoned the trees and, perhaps circling the area once or twice, soon came down into the marsh to sit out our departure at a distance of several hundred yards. The best they could let out was a few croaks.)

It just now occurs to me that there is a question in my mind about the Ospreys' next reaction. When we settled down to photographing Clapper Rails, did they continue their screaming and diving? Did their violent protest continue all day? I can not believe that they failed to calm down, possibly out of sheer weariness, but I am unable to assert that it turned out so. It is possible that we became so absorbed in what we were doing that it was *our* perceptiveness that became weary, while the Ospreys, shut out of our minds, continued to clamor as strenuously as ever.

Whatever the case, I can take up the narrative thread again at the time I climbed to my first nest. At least *those* Ospreys now set up a screaming duet as none of the others had previously. In addition they dived lower, so that as they whooshed past I not only heard the singing of their separated primaries but felt the wind of their passage on my neck. I found this expression of parental anxiety delightful. I realized that I must not keep the birds away from the nest for too long a time—the eggs would quickly become too hot or too cold, depending on the balance between sun and wind—but I allowed myself a number of minutes of indulgence while the Ospreys continued to perform. Now I turned so that I could watch a bird descend from the beginning of its dive. Its straight-on approach was really terrifying, despite my confidence that gentle Ospreys will never actually strike. During its downward glide, that was assisted through a lessening of air resistance as the bird held its wings partly flexed close to its sides, I could see the Osprey simply *enlarging* as it came onward in the same visual field. Its eyes were fixed upon me with a glare that looked like a distillation of raptorial ferocity, and it waggled its head slightly from side to side as if to improve its focus and its aim at my unprotected person. For I was really helpless. The rim of the nest overhung the tree trunk so far that I had to lean backward to get my head up over the edge. Thus, while I obtained a good view of the eggs not far from my nose, I had to hold on tightly with both hands to keep from falling. Had the Ospreys only realized it, my position was vulnerable and they held all the advantages.

I think I must have decided to become an Osprey bander before I left that cedar tree. There certainly was no subsequent era during which I can remember merely wondering about the idea as a possibility. However, owing to my spending the rest of that summer bird-watching on my uncle's sheep ranch in Montana, I was prevented from tackling Avalon's Ospreys until the following year. And then I received an immediate shock. Someone had been there first; all the babies were already banded.

Of course the scientific objective of all this was merely to get the Ospreys banded so that one could begin to learn about their migrations, longevity, and so on. From that standpoint it could not possibly matter who banded them, and I suppose I ought to have been happy to think that, with me as a backup man, the project was now doubly certain to move forward. But let's admit that I was not, in this instance, playing the role of a scientist. The people in the Biological Survey in Washington were supposed to look after that part of it. As one of their "volunteer cooperators," I could serve their purposes without being the least scientific myself, the only requirements I must fulfill being to make infallible identifications of all the birds I reported and not to get the band numbers mixed up. Hence there was no ban on my enjoying the work, and naturally the main reason I had volunteered was because I knew I *would* enjoy it. But I had not foreseen the complication of a competitor. What can one do in such a case? The only rules I knew were two: You must not place a second band on a bird already banded; nor may you remove a band already present to make room for one of your own. To fight and win in this war (for I had immediately acknowledged a state of unremitting hostility), it was necessary simply to *get there first.*

On the date of this dismal revelation I noticed two late nests that still held eggs while ordinary families contained young birds half-fledged or even about to take wing. At once I marked these clutches down as objectives for my opening battle. I would keep them closely under surveillance, in order to band their chicks at the earliest feasible date. When the babies are very small, bands will slip off over their feet, so it is useless to attempt applying such rings until the rapidly growing chicks have attained close to half their adult weight. Thereafter one can affix the bands at any time, and I worked under the assumption that my rival, not knowing his new need for haste, would wait a week or two longer before returning with bands, pliers, and notebook.

That strategy worked as I hoped it would. On August 10 I had the satisfaction of banding two young Ospreys in one nest and three in another. After that, of course, my competitor was bound by the same rules I had been: He could not remove my bands or add his own. Naturally I would not be on hand to observe his chagrin when he arrived to find the tables turned on himself. Nor was I at all keen to gloat over his defeat—that aspect, the need for someone to lose, was simply an unfortunate consequence of there not being enough Osprey nests around to keep two bird-banders happy.

Indeed there were not many birdbanders around, either, so it became inevitable that the two of us should learn each other's identity. When the disclosure came, I found myself in a disadvantageous, or at least an embarrassing, position, for while no bander has any legal rights to particular working areas or bird-banding specialties, it is generally possible to establish a reputation in a given project that other banders will respect by staying away from it. But I, a mere college student, had encroached on the particular hobby of an established Philadelphia businessman. Mr. John A. Gillespie was well known as a leading member of the Delaware Valley Ornithological Club and the Eastern Bird Banding Association. I was so newly enrolled in both organizations that I had not yet learned the boundary markers of the members' several avocations, and I thus felt quite a pang on realizing that the unknown bander of Avalon should be someone as important as the august John Gillespie.

But a pang of what? Certainly, having made a beginning, I was now in no mind to retire. The closest I could come to defining my reactions was that our rivalry was "a pity" (and *that* solved nothing). I fear that I simply began to plot in earnest against John, realizing that henceforth I must canvass Osprey nests at Avalon early in the summer and then make frequent trips to the shore to keep abreast with hatching and growth of young in every tree so that I could nab them the moment their feet and legs became husky enough to retain a band. Let me state unequivocally that throughout this affair John was the gentleman. Never did he say anything that indicated wounded feelings or a sense that I had infringed as an upstart on terrain that he had already preempted. It was unnecessary for him to have given any such signs, however, because I was painfully aware of being a cad and of his knowing how I must feel. But I guess part of being a cad is the inability to stop doing what cads do, so I plotted and schemed like a demon.

Yet there are delays before this drama can continue. In 1930 I spent the summer in Honduras, collecting birds for the Academy of Natural Sciences in Philadelphia, while in 1931 Merida and I spent July and August on our honeymoon in England, where her uncles helped me identify British birds through the knowledge they had gained as bird-nesting boys. (Apparently *all* country lads in England had egg collections in those times, so that I wondered how the various buntings and such had survived.)

Hence it was not until three years after banding those first five
baby Ospreys that I set foot again in John Gillespie's territory.

By this time, however, I was possessed of an additional in-
centive. Of those five young Ospreys, four had made history; that
is, I had received recovery reports of them from the Biological
Survey. Eighty percent was a tremendous yield! What birdbander could resist dabbling further in Ospreys when results of
that degree were promised? When I levelled off from such a pitch
of elation to a more sensible appraisal, I had to admit that four
out of five was well beyond the average recovery rate, even for
Ospreys, and moreover that of those four recoveries, two were
actually valueless. The latter were simply young birds that had
been found and killed, respectively, in Avalon within a few weeks
of my having banded them. In short, they had never migrated
(the found one had a broken wing and probably died), and
they came to their ends while still learning to fly. They thus
added nothing to scientific lore, for the truism that fledglings
suffer a high mortality rate hardly needed strengthening.

On the other hand, that still left the other two out of the five
Ospreys that *had* done something noteworthy, and one of them
had indeed been a spectacular performer. The less flamboyant
Osprey had merely begun its southward migration in 1929 and
been killed very soon in West Virginia (an important place to
die, however, as I shall shortly explain). The last bird also set out
to the south in 1929 and managed to be found in September in
Rosman, North Carolina. But in this case it was picked up with
a broken wing by a saintly W. E. Galloway who apparently
nursed it back to health and released it. For after concealing itself from the avian FBI during another two years, it was suddenly caught again, this time in Florida, on January, 2, 1932, by
George McNaught of Rockledge. I wrote to this gentleman to
learn the details of that episode and heard from him as follows:

"The Biological Survey told me nothing about the bird
being found before I found it. Yew [sic] asked me about
the behavior and condition of the Osprey.
"The bird seemed very savage but he could fly and swim
as well as if he had never been hurt. I caught the bird

in a small steel trap padded with inner tube rubber, which did not hurt the bird in the least. The bird was only captive over night.

Yours very truly,
George McNaught
"The bird has now about five foot of wingspread."

I won't comment about the hawk's alleged swimming ability. On a higher plane one can conclude that this Osprey had flown south from a northern summer more than once before coming into human hands again in Florida. So let's band a lot more Ospreys if they will reveal their adventurous travels like this!

Exhortations may be good for the soul, but they don't really put bands on birds. Perhaps it was John Gillespie indulging in some counterexhortations, or even a bit of plotting and scheming on his own part to neutralize my brand of witchery. Whatever the truth might be on his side, I can claim to have put a strong effort into my 1932 program but to have met repeatedly with baby Ospreys already wearing Gillespie bands. My final score was not a great improvement over 1929: I had banded only eight birds in five nests.

But at the last moment—again in August—I had found a strategem that should really make me a winner in 1933. It was based simply on two questions: Why stick to Cedar Island and Avalon? What was wrong with other Osprey nests all over Cape May County? Formerly I had often seen such nests, both along the mainland border of the salt marshes, and actually inland, although never more than about two miles from the ocean or Delaware Bay. But I had dismissed those nests for being either in trees that I could not climb or else on private property that I could not invade. Trees were actually the more important factor, for none of them had convenient staircase lower branches like Avalon's cedars. Yet there was such an abundance of Ospreys' nests to be seen in a drive along roads in the lower part of the county that one should be able to find some climbable ones among them —perhaps enough to keep me busy and happy. My last three 1932 birds had indeed been hatched in two nests situated not in Avalon but near Cape May Court House. Therefore 1933 should wit-

ness a shift in my onslaught, during which I could *almost* leave
Avalon honorably to its first banding devotee (but only almost).

It was already clear to me, however, that unless I distinctly
improved in my climbing ability, the increase in hunting range
would not give me the great boost in banding performance that
I coveted. The world's worst athlete must give up reliance on
spidery arms and legs alone to become a professional climber
with leg irons, belts, ropes, pulleys, and all the rest of such gear
that falconers use to reach eyries of eagles and peregrines. Probably my best course would have been to go to a school for telephone linemen. Another thing I ought to have done—and this
really might have solved all problems—is to have found a partner,
for climbing with ropes can be accomplished much more easily
by teams than by hermits. But I quit almost before I got started,
for a reason that seemed valid enough at the time and still is, as
far as I can think into the question. I bought a pair of climbing
irons and early in the spring of 1933 took them to South Jersey
for a practice session on the trunks of large pine trees. At the very
first step I attempted to take into the sky, I fell back to earth because I was unable to dig the spike of the climbing iron into the
wood of the tree. Instead it penetrated only into superficial,
flakey bark layers and then ripped out as soon as I put some
weight on that leg. I then tried a couple of dead trees, but their
rotted wood was too soft to hold the spike solidly—at least I felt
that I might slip at any moment. Thus I gave up that approach,
with the reflection that telephone linemen have it pretty soft with
their beautifully peeled and chemically treated poles. It is barely
possible, I must admit notwithstanding, that someone with greater
athletic ability than mine might not have failed as I did. A more
muscular and compact person could have dug the spikes in
harder, but he would also probably be heavier and more at risk
in dead trees and on dead branches of live ones (where I was
often relatively safe), so it is difficult to know whether my defeat was entirely personal or was generally ordained by the trees.

Thus I returned to the project with no more than my original armament, though since that included an overflowing store of
enthusiasm and enough foolhardiness to have done in an acci-
dent-prone individual, the auspices were not as bad as I had

feared when the climbing irons petered out. Moreover, Merida and I spent that summer in Ocean City, which was only a few miles up the coast and, short of being in Avalon itself, about as close to the area as I need be for crucial infighting. For I planned to get up long before dawn most of the time, to be in lower parts of the county at daybreak, and still to get back to Ocean City in time to find more normal people finishing late breakfasts. I was, incidentally, a medical student by this time, but even in those days I was attracted to the healing field as a science rather than an art, and I would not have dreamed of exchanging a summer of Osprey ornithology for the sorts of junior internships that many of my classmates were adoringly sweating out.

By this time, in fact, the Osprey project had taken on a greater semblance of science than it had when I was banding merely for the personal joy of it in blind confidence that the effort was doing general good for the Biological Survey. John Gillespie had developed a theory about the first southward migration of young Ospreys, and to test out his views, the importance of banding as many young birds as possible had now increased. From now on it really did not matter who banded the birds, though John and I would both keep at it as hard as ever. He may have recognized my zeal, but if so he refused to let it lead to personal rivalry. Rather, he wholly approved my expanded ambitions, which I had disclosed to him. Moreover, it seemed that he had no intentions of working outside of Avalon, so I would have no competition elsewhere.

John Gillespie had now banded dozens of young Ospreys, and while dozens of other sorts of birds, such as sparrows, would be considered only a few, that number of this giant species was a very great lot indeed. It had taken him many years to build up his proud total, and consequently his tally of recovery reports from the Biological Survey had become similarly impressive. His birds had all been banded as nestlings on the New Jersey coast, so that in an unpremeditated way he had set up a splendid investigation of the behavior of young Ospreys on their first southward migration in late summer and early fall.

The Fish Hawk, or Osprey, does not confine its breeding to seacoasts. Apparently the species finds trout as acceptable as any saltwater fish. Consequently its nests may be found also along the larger inland rivers and lakes of the world. Thus ornithologists had never found Ospreys out of place at Hawk Mountain, Pennsylvania, for instance, where the birds join other hawks in traveling southwest along the Appalachian mountain system during September and October. But these individuals were generally believed to have bred in the interior somewhere, while coastal Ospreys would equally readily be presumed to follow the waterline southward as their landmark.

However, as John's data gradually accumulated, he became convinced that this presumption must be modified. One after another of his birds was found—trapped, shot, or otherwise encountered—in the mountains only a few weeks after it had learned to fly in Avalon. These recovery reports did not include Hawk Mountain, Pennsylvania, which would be somewhat to the north, but they mentioned almost every state to the south, beginning with Virginia. The birds had apparently set out in a westerly direction until they intersected mountain ridges that would then guide them ultimately to the Gulf Coast.

On the other hand, a few of Gillespie's surviving Ospreys, ones that happened to meet tragedy during the fall migration of their second, third, or later years, came to their ends in coastal regions along the flyway that one would have considered the correct one in the first place. There weren't quite enough recoveries in this category to allow one to generalize confidently on the subject, but John did come up, nevertheless, with an interpretation that sounded plausible to the non-ospreyan human mind and that demanded much further Osprey banding to confirm or negate the conjecture.

It would not be fair to invite you to watch a young Osprey as it learns to fly and then to catch its first fish, for you are not likely to have that chance. Therefore you must indulge me when I report that those procedures are prolonged ones, filled with awkward false starts, mistakes, and failures, so that it is a miracle that the fledglings actually master those arts in the few weeks of practice available to them before they must set out on a flight of one or two thousand miles.

The crux of John's theory was that the birds were indeed

still inept fishermen when the necessary time for them to move arrived. Therefore he proposed that they had better go where the fishing is easiest. Obviously coastal waters may be extremely turbulent during autumn storms, while even in relatively calm periods the open stretches of bays and estuaries are never fully still. Would it not be much safer—or simpler—to hover over the surface of a pond or slow stream within the shelter of wind-subduing mountain shoulders? Here the smooth liquid face would not give deceptive jumpiness to a fish's shape, so that one's as yet inexperienced aim would have some chance of success after the long, sliding aerial dive. Moreoever one could also more easily burst back into the air, carrying the flopping burden, than if it were necessary to contend with a running surf as well.

The demonstration of a consistent migratory pattern, in which youngsters flew south by one route and oldsters by another, would be nothing new to the ornithological world, though as regards Ospreys such a disclosure would be heralded, like any other new scientific fact, with general acclaim and with special plaudits for its innovator. Not that John Gillespie was interested in fame—he cared much more directly for Ospreys alone, and his curiosity about their doings occupied a detached portion of his mind that maintained no neural connections with whatever underdeveloped bump of vanity nature may have given him. But for the theory to stand, it must be supported by massive data— a number of recovery reports from the Biological Survey far greater than he himself could ever hope to receive from such additional dozens of Ospreys as he would still have the opportunity to band in his active lifetime. So that is why he welcomed my arrival in the Avalon arena. To his satisfaction, as well as to my own, we became Osprey collaborators rather than competitors.

Thus I put on a really big push in 1933, banding a thumping fifty-seven Ospreys in twenty-nine nests. Only two of the nests were in Avalon, and those were late ones that John must have missed. The others were largely near Cape May Court House, though a few localities, such as Dias Creek and Green Creek, lay across the peninsula on the Delaware Bay side. Nests near marshes were mainly in cedar trees, while in drier parts, pitch pines were the most popular supports. Wild cherries held seven nests, oaks three, unidentified dead trees two, and one each was on a pole and a platform. That does not necessarily indicate the Ospreys' preferred order of nesting sites, however, since the tabulation lists only the ones I was able to reach. Suppose there had been a hundred nests on inaccessible chimneys: My banding records would not have mentioned these and would therefore have been misleading to anyone who tried to use them as a guide to Ospreys' nesting habits. Actually I did not see any chimney nests, but there were a great many that I had to omit because they were equally unattainable. In some cases the trees were simply too big and I was afraid to climb them. In others they possessed too few side branches for footholds and handholds. Some trees were obviously rotted nearly through; they could only just support the huge weight of the great nest, and the addition of my 125 pounds would bring the risk of snapping the trunks almost beyond an even chance. A few nests were built in places where I could not venture—islands, private property, telephone poles, and so on.

I could now kick myself for not having made a list of all the nests I failed to include in the project, for in this day of sharply declining Osprey populations, such negative data, added to my list of successes, would give a revealing retrospective picture of how common the birds were in 1933. From memory I should say that I must have banded young birds in less than half the nests I discovered. Since I canvassed the main highways of Cape May County rather thoroughly that year, it might be safe to say that I knew of a hundred easily found nests. John Gillespie probably kept another 15 nests in Avalon under surveillance, bringing the known total a bit higher. It is useless to wonder how many undiscovered nests there were in wilder areas, for the question is unanswerable. But I would not argue with anyone who wanted to support a conjecture that between 250 and 500 pairs of Ospreys used to breed in Cape May County in the early 1930s.

Despite the bruhaha I have raised about my entry into the science of baby Osprey migration, I resided only briefly in the study. John Gillespie finally acquired an assistant, such as I had dreamed of, and at the end of sixteen years had banded a total of 457 fledgelings, which made my 70 seem a small contribution. He heard from 76 of his, and I from 14 of mine. Thus there were only 90 records from which to sift the information we wanted. But some of these referred to birds that were killed or found for other reasons before they had migrated, while many pertained to individuals already on the wintering grounds, so that one could not surmise which southward route they had taken. And of course we heard also from older birds at a variety of times and in various places, but these did not contribute to the theory either. In all, no more than 38 of John's birds and 8 of mine could qualify for analysis.

It seems safe to conclude that a fairly large proportion of recently fledged Ospreys *does* use the inland flyway. But our records showed that the rest of them followed the coastline. What can one infer from such data? That biological systems are maddeningly irregular? That's true enough, but so far as Osprey migration is concerned, we really could not say much more than the obvious—that the young birds seemed to scatter, so that some wandered to the interior and therefore had to find their way south by the overland route, while others happened to remain near the ocean. As for our further conjectures, that older birds shunned the interior in fall but wisely took the easy coastal path to Palm Beach and Havana, we simply did not have enough recovery reports either to support or demolish such proposals.

Now that Ospreys have become so much less common—almost rare—allegedly because of concentrations of pesticides in their tissues, I find it poignant to think of that summer, close to forty years ago, when I lived so many early mornings with them. It would be a mistake to say that I established any sort of intimacy with the birds. As a matter of truth, I ended that season with a wholly modified view of Ospreys' supposed docility. I explained this change of attitude to Merida on a July morning in Ocean City while she sponged caked blood from slash wounds in my scalp and one of my ears.

I had found a rather unusual nest near Cape May Court House. From the road it was barely visible near the top of a large pitch pine that was situated at the center of a thickly wooded swamp at the far side of a broad field. As I crossed the field, I had lost sight of the tree, but I kept in as straight a line through the woods as tangles of catbrier would allow and eventually looked up at the nest. This was a lot higher than I had imagined; however, there were a few sketchy footholds, and I was presently rather breathlessly peering over the rim at three very young fledgelings—old enough to band, fortunately, but nevertheless at a stage when down and pinfeathers, rather than expanded plumes, remained their nursery garb.

There are times when I weep for a spider monkey's tail. This was one of them. You must visualize my position. My feet were on a six-inch stub at the trunk of the tree. From there my legs and torso angled out at about thirty degrees in order for my head to clear the body of the nest and for my eyes to look over its edge. My two hands consequently held on as a lifesaving measure. Granted I could have clung with one hand, the other was not by itself dexterous enough to manipulate pliers, bands, and baby Ospreys all at once. A tail would have solved everything. Yet the job had to be done. Chins are poor substitutes for prehensile tails, but I found that if I hooked my jaw into the nest and pressed my elbows against its bottom, I could just barely make the necessary manual moves.

Nor did the babies help. As a rule, birds of their age cower and let themselves be handled without resisting. These youngsters chose to be defensive. They reared back and then struck with their soft, warm beaks in a ridiculous way that totally failed to look intimidating but which nonetheless increased my difficulties.

This must have been a savage breed of Osprey, the scions precociously reacting to danger with such violence as they could generate. I do not remember being struck. I was suddenly simply holding on, knowing that I must get down at once. Something warm was dripping from my nose. I seemed to have a tremendous headache and a stunned deafness, as after a clap on the ears. I *did* get down somehow—and it must have been fast, for just as I ducked below the nest, an Osprey swept past on its second sally, wafting me with its wings. Later I found that I had instinctively retrieved my bands and pliers, for they had been in the nest and here they were in my pocket. However, for the time being I was "punchy," not quite sure of anything. I could not find my way out of the woods, actually circling (for the only time in my life) and returning several times to the base of the tree.

Besides disproving my notion that Ospreys will never attack, these birds had furnished me with the baseline for a splendid scientific study of Osprey behavior. After an additional few decades of exposure, I would be in a position to report what percentage of the species is wild enough to knock you out of trees. But there was—and I suppose there still is—something craven in my morality. I simply did not indulge that opportunity for unbiased research. Frankly, I lost my nerve and gave up Osprey banding after that season.

Yet I did not need to band Ospreys to have communion with them. After a hurricane (we called it a nor'easter) in Avalon one summer, I sloshed across the still-inundated salt meadows to see what havoc could be tallied for that particular storm. At the edge of a minor cedar island that no longer exists I came upon an upside-down Ospreys' nest. Its supporting dead tree had been leveled, while the nest itself lay completely overturned in the water. I tugged at it, but could not have budged the mass even if it were not sodden. No Osprey could be seen or heard in the sky. I concluded that whatever life the nest may have held had left it before the blow. As I was about to leave, my eye detected a small movement in the debris, and when I tore away a couple of wiry cedar branches I was amazed to find first a foot, then a living fledgeling Osprey's leg complete with Biological Survey band (which, naturally, had been put there by John Gillespie).

Now I had to dissect the nest piece by piece, taking care not to inflict further injury on a bird that must be already near death. Such work of disassembling was a tedious job, since I was delving through the bottom. As I reflect on it now, I realize that the nest must have been badly smashed, or else the foot would not have been showing, nor could I have penetrated without an axe or a blowtorch. But whatever advantages happened luckily to be on my side were sufficient to enable me at last to hold up a feeble and totally disheveled Osprey about two-thirds grown.

My return to Ocean City that morning was more in the nature of a triumph, since instead of being injured I bore a trophy. Merida may have been no less dismayed, however, when I introduced the mud-caked new member of the household. "He" (I never knew the bird's sex) became "Ozzie" at once. I rushed to a nearby bait shop which blessedly had a boundless supply of live minnows. Consequently I faced no feeding problem whatsoever— there was no need to teach Ozzie to eat some esoteric kind of processed fish, such as canned catfood. The moment I returned and showed him a minnow wriggling in my fingers, his drooping head shot up open-beaked, and in a short time I had him crammed full.

That was on August 25. Labor Day swept by only a few days later, or so it seemed. We moved to West Philadelphia while I went to Penn Medical School that fall, but as this was no place to complete rearing an Osprey, I carried him to my parents' home in suburban St. Davids. There I erected an artificial nest in the vegetable garden, mounting some chickenwire and wooden crosspieces at the end of a pole and dumping most of the compost heap on top of that foundation. It really did look something like a nest and Ozzie, for one, accepted it as such at once. Like the future Elsa, who was "born free," Ozzie was going to have his chance to return to the wild if I could provide it for him. From the time I heaved him up into his new home, I never put a finger on him again. I would toss fish into the nest and it would then be up to him to handle it. As long as he remained there, I would continue feeding him, but my hope was that he should someday disappear without a trace (alternative to being found grounded—injured or dead—in the neighborhood).

Birds being creatures largely of instinct rather than reason, it was difficult for me to interpret Ozzie's responses to his new situation. Perhaps the trouble lay in my looking for any responses at all, whereas Ozzie possibly behaved as he did oblivious to his surroundings. If he had still been in the family nest at Avalon, would he not have exercised his developing wings exactly as he did here in St. Davids? Would he not have spent just as much time studying a sky that looked mostly vacant to me but occupied his attention as if it were trooping with jostling flocks? The only real difference that may have affected Ozzie was that he was now fed by a grotesque biped that stood below him on the ground, rather than at the arrival of an aerial food conveyer.

(Also he was now feasting on swordfish steaks, there being no bait shops close to St. Davids). I think he did find me odd. Whenever I came near, he would waggle his head, keeping me transfixed in his binocular gaze, but by the shift of his visual field causing my image to oscillate on his retinae. In this way he must have improved his "fix" on my exact shape. He looked as if he was straining to make me out as another Osprey, or, failing that, to figure out what sort of non-Osprey I might be.

In any event it was not long (about mid-September) before he took his first flight. By chance I was there to see the performance. It was a blustery day, and Ozzie had been standing, spread-winged, in the nest, catching gusts that would lift him up a foot or two before he settled down again. With one particular gust he suddenly blew clear of the foundation. Now he must flap with all his might to avoid a fall. And that suddenly launched him at a speed that even he must not have foreseen, because in a moment he almost crashed into the side of the garage. With even more strenuous flapping he avoided that obstacle, only to find himself headed toward another. The backyard, which I had always considered spacious, was almost like a cage to this gigantic bird. Here was an additional reason why Ospreys do well to nest in elevated, exposed situations: They need a large, empty arena in which to earn pilots' licenses.

With a final great outburst of flapping, Ozzie cleared the hemming treetops and disappeared. That, I concluded, was that —and not too bad an ending to the story. However, I threw some fish into the nest anyhow, just in case he might come back. The next day (I was making daily visits from West Philadelphia) I found the nest empty and yesterday's untouched fish still there. Such is the condition of our senses that they sometimes play tricks of association on us. Though I could see no Osprey, I was looking at an Osprey's nest and I was thinking about Ospreys: *Ergo*, I thought I heard such a bird's high, broken whistling call. The illusion became so strong that I could no longer believe it was only a fancy. And at last I found him.

Our neighbor, Charles H. Stewart, was vice-president of the American Radio Relay League, a pioneering society of radio amateurs or "hams." Because of his distinguished office, he maintained an especially powerful station (3ZS). In those days you could receive and send the farthest according to the height of your antenna. Mr. Stewart had scooped lesser hams by having the telephone company erect an Ossa-on-Pelion type of rig—a double pole, the base of one clamped to the top of another. There sat Ozzie at the pinnacle of that perfectly designed Osprey perch, screeching and waggling his head at me as if I were a long-lost friend.

I quickly threw a fresh piece of fish into the nest. Though he seemed excited—and I am certain he could see the food and distinguish it for what it was—he would not come down. That *was* a narrow vortex for an inexperienced flier to descend, while moreover he still had a fresh recollection of his difficulty in rising out of it. I think his screaming was supposed to guide me in my fish-dispensing: "Not in the nest, Stupid! I'm up here!" Meanwhile his antics were becoming a cause for wonder to other birds in the vicinity. I saw both Sharp-shinned and Cooper's Hawks fly around him, clearly undecided whether to strike or maintain cautious inaction. Ozzie was not only alien to the locale, but his behavior made him grossly suspect as well.

I know that he made it at least once (though no one had been there to see how he managed it), for a large piece of fish was missing, whereas the nest was proof against raccoons, cats and opossums that might attempt to maraud it from below. Thereafter on one of my visits he would be absent, but then next time he might be back as vice-presidential emblem for 3ZS. I saw him last on September 20. However, although that was by no means too early a date for Ospreys to begin migrating, Ozzie did not set out yet. My mother was told by one of her friends, Mrs. Georgia Walton, that a huge bird had arrived out of nowhere and was systematically catching all the prize goldfish in her skating pond. "Why, that must be Brooke's Osprey," said Mother. Such was the lady's greatness that she immediately called off plans that had been set for the bird's elimination.

I reported Ozzie's history, together with his band number, A721063, to the Biological Survey. Since that number had not been issued to me but to John Gillespie, I would never receive further word of Ozzie from the government but must depend on

John to tell me if any news ever came to him. And so, indeed, it did. According to the official entry in Washington, Ozzie's story terminated with the facts that a leg and band were found on the following first of May at Edenton, North Carolina. John gave me the finder's name as well as the other data he had received, and my letter to the southern gentleman concerned elicited the information that the bird had been killed the previous fall. The man, a farmer, had seen it at that time, lying at the far side of his field, but thought it was a dead rabbit. Only when working on that side after the passage of winter had he come close enough to realize that the carcass had a covering of feathers rather than fur.

So what was Ozzie's final message? For one thing, he had flown 250 miles from St. Davids and had been going in the right direction. So far as his being restored to a wild state is concerned, I am claiming that Ozzie had been rehabilitated. Certainly many of Avalon's young Ospreys don't even make it out of Cape May County before meeting destruction. Whether he got all the way to Edenton on Mrs. Walton's goldfish is an open question. In any case, he had learned to fly and to fish all on his own. A wry reflection is that Edenton lies on the coast, at the head of Albemarle Sound, so that Ozzie was flying south against "Gillespie's Theory." But as he had started from the wrong place, I am in favor of scratching his contribution from the statistical table for the likelihood of its being atypical or erratic.

The end of Osprey banding by no means stopped my visits to Cedar Island. I had my eye on the herons nesting there. (My God! I've still got Victor Liguori holding a baby Snowy Egret for me! But never mind, I'll rescue him soon.) Although heron banding may be revolting, as I have already detailed, it is at least safe. While my permit was new, I wanted to band anything and everything, just to make the quantities as large as possible. But in addition to being simply additional grist, the herons had their valid intrinsic interest. And being relatively large birds, they attracted me further by the increased likelihood that eventual recovery reports from the Biological Survey would be forthcoming in fair numbers.

I no longer swam to the island in secret. Rather, I now hired a boat from a real old salt whose sign, "C. E. Fort," was the only indication that he was in public business. When he found that I did not want to go farther than across the channel, or about three strokes of the oars, he demurred about accepting a fee, and I think we finally settled for twenty-five cents for the whole morning. I visited there irregularly, but for ten years or so, whereby I ultimately became well acquainted with the heronry as well as old Mr. Fort. Most of the birds were Black-crowned Night Herons, though I located a few families of Green Herons and knew there must be one or two of Yellow-crowned Night Herons too. Initially the colony was a small one: Perhaps its maximum strength consisted only of fifty pairs of birds. In retrospect that now seems like a lot.

My first effort led to a goose egg. It was too late in the season and the babies were already climbing like monkeys as well as flying a bit. The next year I came at once upon catastrophe. In the middle of the heronry, in a small clearing under a large holly tree, I found signs of a fire, while nearby lay a blood-soaked heap of skins, feathers, bones and entrails, and pinfeathers indicating that all had come from young birds. I could not find any babies to band, and the adult herons stayed well out in the marsh, away from their raped incubator. I quizzed Mr. Fort who enlightened me at once. "Some of the young men from the Coast Guard came over and had a barbecue last Saturday night."

Here was one branch of the government fouling the work of another. Herons are of course on the list of permanently protected birds, but even if they were not, we have no game species that becomes vulnerable at the squab stage. I wrote agitatedly to the Biological Survey. After a long lapse, during which I imagined every kind of penalty or fine that might be visited on the fiends, I heard from Washington, and the answer has always impressed me as the most Solomonesque in my experience. "We have designated the Avalon Coast Guard special custodians of the Cedar Island heron colony."

After that the heronry nevertheless declined for a number of years. By 1940 it had built up fairly well once again and I—of all people—became an agent in its being disrupted a second time. During that year I was at the Rockefeller Institute for Medical Research near Princeton, studying a virus that causes foot-pox in Slate-colored Juncos. One of my new colleagues was Dr. Malcolm S. Ferguson, a parasitologist at the time specializing in trematodes. If pox viruses of Juncos sound far out, you should hear what Fergie had just come up with. One of the New Jersey State fish hatcheries had had some trouble with its fish dying, and Fergie had been called to investigate. The only unusual event that he could unearth was that earlier in the year the ponds had been visited by some seagulls. By the time Fergie pieced the entire story together, it came out as follows. The gulls had been infected with intestinal trematodes, or flukes if you'd rather. The flukes' eggs were discharged into the water with the gulls' droppings. The eggs hatched into swimming larvae that invaded snails, parasitizing them. Some time later a secondary type of trematode larva emerged from the snails but now invaded the young fish of the hatchery. Here they were drawn uncannily to the lenses of the fishes' eyes into which they penetrated, causing inflammation, destruction, and finally opacity. The fish being blinded starved to death and were soon floating belly up as invitations to be eaten by other seagulls. Had such birds arrived again, they would have digested all but the second-stage trematode larvae. These would have been chemically liberated from the eyeballs and would now mature to the adult stage in the gulls' intestines.

When Fergie told me he needed some young herons to test out the suspected life cycle of another kind of trematode, I assured him immediately that, he having all the necessary federal and state permits, I would take him to Avalon on a collecting expedition. By this time the rookery, in addition to becoming more populous, was being watched to see when it would add the Little Blue Heron to its breeding roster. That species had formerly bred along the South Jersey coast, but toward the end of the last century it had been reduced in numbers along with the egrets and had only rather recently come back. As Fergie and I trudged across the marsh with his cages, three white birds, which I carelessly identified as Little Blues, flew out from the line of cedars ahead of us. But then, in a double take, I remembered that only the immature Little Blue Heron is white—the adults are blue as their name proclaims. Since May 29 was much too early for young Little Blues of that season to be on the wing, perhaps I had seen Snowy Egrets. Well! That would be a much more notable "first," for Snowies had been almost entirely exterminated and had, indeed, been at dead center of the entire controversy over feathers in ladies' hats.

The three birds had settled in the open marsh. To be sure, they *were* Snowies, with their strikingly yellow feet at the ends of black legs. Now it was still a question whether they were actually nesting here. Fergie and I soon found a nest that I felt sure must belong, if not to the trio, to two of the adults. It contained only a single egg. However, among the Night Heron nests and eggs, this exhibit stood out for its smallness and greater delicacy. Color was no help; all the eggs were a "pale, dull blue," and if there had been any Little Blue Herons present, it would have been more of the same. Not seeing any Little Blues at all that day, I confidently recorded the nest as a "Snowy's, though one's confidence might be of a more comfortable sort if one could follow up the matter by spending a vigil in concealment and seeing a Snowy Egret return to squat on that egg.

But we were here on Fergie's business and did not have time to play my bird-watching games. The rookery was really doing quite well. I estimated there were about twenty Night Herons' nests. Fergie collected seven very small (nonclimbing) babies as well as ten eggs that he would hatch out in an incubator at the Institute. I marked the position of the egrets' nest in my mind so that I could find it on our next visit.

On June 18 there were six eggs. I saw only one adult egret, but that one remained in a tree close by, obviously beginning to feel the parental urge at this late stage of embryonic growth. On July 8 we saw two adults and the nest contained four babies, all of which I banded. At last I could aver that the progeny were those of Snowy Egrets, for the tips of their four outer primaries lacked dark tips (as I had lectured Victor L.). I don't know what had happened to the other two eggs, but four youngsters was a thoroughly satisfactory crop.

Now I had a problem. This was banner headline news, but I could not publish it freely as such. If I blabbed all the details of the story, Cedar Island would be overrun by egg collectors next year, and the "Snowy Egrets' comeback would fall flat. Yet I wanted to inform my more trustworthy fellow ornithologists, and with that undeniable grain of vanity that I possess asserting itself, I wanted to receive personal credit for the discovery. Today, these three decades later, I am not able to tell exactly what I did about it. It was bound to have been something secretive, but so well did I hide the news that I can now find no trace of its outlet in my records. And of course the "eggers" were foiled, since the Snowy Egret is currently a common breeding bird in Cape May County.

Victor said he had been terrified during the whole of my absence, which had seemed endless though it was only a few minutes. (These reminiscences also flashed by in my mind at high speed: I honestly did not dawdle.) He had never held such a bird before and had no idea what it might do to him all of a sudden. He said its eyes had looked at him most evilly. Poor thing— it was frightened to death, too, so why shouldn't it look livid? Anyhow, my banding of it brought my Cedar Island Snowy Egret grand total to five, a figure that still stands.

As we walked farther that day, I was saddened to see how few of the cedars held tenanted Ospreys' nests and wondered how long it would be before the whole island fell into some developer's scheme. But less than a year later, a most wonderful thing happened—something that still seems almost impossible to me. In the *Philadelphia Evening Bulletin* of April 10, 1968, I suddenly fastened my vision on an electrifying headline: "Avalong to Get $135,000 for Cedar Island." Hardly caring to read further (for what difference did it make which real estate firm had swung the juicy deal?), I glanced down the column nevertheless and found that the Borough of Avalon had in fact sold the 190-acre island to the State Department of Conservation and Economic Development under its so-called Green Acres program. In effect this meant that Cedar Island would be even better off than it had been formerly, for now it could be classified as an official wildlife area, with its flora and fauna under a protective custody that could no longer be threatened.

And they lived happily ever after—ha ha. That summer a man of twenty-six and two teen-agers were apprehended on Cedar Island after they had thoroughly upset the breeding Ospreys by using the birds as targets for bow and arrow shooting.

Chapter 3

LUNA MOTHS

The first intimation I had of the presence of luna moths on my farm came on August 12, 1966, when I found a green left-front wing, broadly margined with purple along its front border, lying in the grass in my incipient orchard. I remember thinking then that the wing showed remarkably little wear for that date, because lunas should have emerged from their cocoons, mated, laid eggs and died two months earlier, and any remains floating about on breezes since that time ought now to be almost too shabby for recognition. What a little I knew about luna moths in Cape May County on that occasion!

In fact, I knew very little about luna moths anywhere, for this particular large saturniid gem was a species that had always kept itself scarce along my journeys into the wilds; or perhaps my wanderings had not taken me to the right places. I thought of it as an inhabitant of deep forests, especially in the mountains, for as a boy I had seen it come to lights at night in the Catskills, while later I had collected specimens, also at lights, when I served as a nature counselor at a boys' camp in Vermont. However, I remember seeing one unattainably high on a telephone pole under an arc street lamp in St. Davids, Pennsylvania, which is only about ten miles from Philadelphia, while the first one I ever encountered lived in Coatesville, not many miles farther west. Thus New Jersey flatlands and second growth woods, though not exactly my idea of luna country, might not be altogether hostile or alien to those moths.

That suspicion proved to be fact. On June 18 of the following year I caught a rather battered female luna moth in one of my bird nets. After carrying her back to the house, I placed her in a cardboard box, with a sprig of persimmon leaves to make her feel at home and actually to *invite* her to lay eggs, for it is known that some moths at times will die with packed abdomens rather than oviposit on substrates that obviously are unsuitable as future larval food. My female was not as decrepit as I had suspected her of being. At least she carried herself extremely well, considering (as I must) that she was still very much pregnant. Within two days she had laid 207 eggs, singly, in short rows, or in small clusters, some on the leaves, but most of them on the sides and bottom of the box. Feeling that that number was more than adequate for any games I wanted to play, I gratefully gave the moth the brief freedom she could still claim and then turned my attention to pediatric chores.

The first thing to do was to get the eggs (which under a lens looked beautifully marbled, cream, brown, and fawn, like plovers' eggs) as quickly as possible into a more natural environment. The interior of a box, inside a house, was likely to be too desiccating for them, despite the presence of leaves. For the foliage was now wilted, testimony enough that although still attached to a twig, it was constantly breathing out moisture. When we deal with something as small as a caterpillar's egg, it is necessary to bring our sights down to the proportions of its world. A persimmon leaf then becomes a rather prodigious structure, upon which a single egg could easily be overlooked. The importance of this observation is that the egg exists not so much in the environment occupied by the leaf as in the environment of the leaf itself. In other words, the leaf manufactures its own immediate microclimate that, within a few hairsbreadths of its surface, may differ markedly from that of the circumambient atmosphere. Therefore if we were to measure the humidity of the air under a persimmon tree at noon on an average sunny day in early summer, we would probably get a much lower reading than that which would be true of the film of air touching and immediately adjacent to each green leaf. As a corrollary of these particulars, I have often thought that a caterpillar must do more than merely eat leaves. As it munches away, it constantly bites into fresh living tissue which must "bleed" in consequence. Does not the caterpillar then feast and *drink* flowing sap from the tree at the same time? I have not yet decided how to test that supposition, though it might be possible to determine whether, in eating a single leaf, the larva gained more weight than the leaf alone could provide.

So far as the eggs were concerned, I could not now give them a perfect home, but my proposed substitute would be adequate, as I knew from past experience. The luna had glued each egg to a plucked leaf or to the unbreathing cardboard surface, and I could not very well unglue them and then reglue each one onto a living spray. Aside from that being a terribly tedious and delicate task, I did not know where to get waterproof glue of a quality equal to that of the luna moth. Just consider the virtues of that cement: besides being quick-drying, long-lasting, and insoluble in water after having set, it had a liquid base whose volatile principle was completely nontoxic to living tissue of animal and plant. A droplet of this admirable glue was exuded with each egg in such a fashion that the moth's ovipositor simultaneously daubed the receiving surface with the viscous black fluid and placed an egg in its middle. And as for its waterproof qualities, I should mention that I have seen last year's caterpillar eggshells still adhering to twigs where they had resisted four seasons of snow and rain.

It did not take me long to cut the box into fragments. I threaded the pieces that bore one or more eggs on a piece of string, also tying in the two or three egg-bearing leaves; and then the entire beaded necklace could be fastened closely among living leaves at the end of a persimmon branch. Perhaps that was crude, but it was the best I could do, and as I have already said, I knew it would work. But whereas this took sufficient care of microclimatic requirements, at least for the duration of embryonic development, it made no provision against more remote external dangers. For while the physiology of larvae within close proximity of leaves is vitally important, caterpillars do not, for all their intimate dependence on foliage, exist in secluded sanctuaries that are accessible only to the microscopist's eye. Now we must again raise our sights to the world at large and see the host of enemies that would gladly feast on or parasitize luna eggs and caterpillars. As in an alien jungle we cringe in terror of unknown dangers that surely must be concealed in depths we cannot see, so I could not name all of luna's foes, not having had prior experience with luna herself or with caterpillar enemies in general in South Jersey. Thus I could only say "birds," and "parasitic wasps and flies," "predacious wasps and bugs" and so on, without naming the kinds and without being certain that they all really lurked ready to pounce. However, a bag of netting, enclosing the persimmon branch with its cardboard polygons of developing eggs, excluded lions and tigers for the time being and assured me an opportunity to follow the next step—provided, of course, that the female had been inseminated, a contingency which now became so urgently to be hoped for that, as days went by, I found it difficult to refrain from dissecting a few eggs in a search for embryos.

All my life has been one long drag, counting the days until Christmas. But then I have enjoyed hundreds and hundreds of Christmases, which is many more than the average person can claim. I felt very certain that the eggs must be fertile, not only because a number of male lunas had preceded my female in the bird net, but also because she had laid eggs in the box with such abandon. Even with persimmon leaves present to encourage her, she might have been more reluctant to oviposit if virginal. Thus all the signs were favorable, and I gave the 207 eggs no hurry-up shots or premature midwifery.

To me it is not enough to "know" a luna moth, gratifying though that knowledge most certainly is. I want to see the insect in all its stages; in fact I must follow it through the egg, through each molt of the caterpillar and through the pupa in the cocoon, before I feel properly acquainted with the emergent winged adult. Therefore to whomsoever finds caterpillars repugnant, I can only protest that I do *not*, and that consequently he must be blind as well as wrong.

But in addition to enjoying the full esthetic display of my captives, I recognized the opportunity to study them more soberly. For example, the most obvious line to pursue (though I would not have thought of it in my teens) had to do with all these eggs' being sisters and brothers. Let me confess at once that I *assumed* them to be full sisters and brothers, the product of a single father as well as only one mother, for saturniid female moths usually accept only one mate. Let me confess also that I might have been wrong on that score: it does not make too much difference, because 207 caterpillars with one mother and with at most two or three fathers shared among them would still make a most interesting family. In other words, I had the chance to study variations, if there were any, among a large batch of siblings. And the sorts of variation that I could detect might be of any kind whatsoever—in form, in behavior, in development, or whatnot. And since I would rear the caterpillars under uniform conditions in the same net, or in nets on the same tree, these observed differences might be catalogued under headings ascribed to genetic variability in the species *Actias luna* in Southern New Jersey. *Should* such lack of uniformity appear, I could state that lunas had different kinds of physical selves or even persons, and perhaps I could actually emphasize some of these traits by further breeding experiments (in which, moreover, I could be certain that individual batches of progeny sprang from a single sire).

How early in their careers might luna caterpillars show up as recognizably different personalities? I had no way of even guessing at the answer, though it occurred to me that the first signs could possibly be manifested at hatching time. What if some larvae chewed their way out of enclosing eggshells sooner than others? That would at once separate them into the go-getters and the lazy ones. But then I remembered—alas!—that they were already all of different ages, because fertilization of an egg is not effected until it is in the process of being laid. Nature has so rigged the insides of female luna moths (and other insects) that they store their mates' sperms in a seminal receptacle, and it is only as an egg passes by the mouth of this reservoir on its way to the outside world that it is fertilized, in the very last nick of time. Therefore no two eggs had begun their embryonic development at the same instant, and they *should*, therefore, hatch at intervals from one another if they grew and formed their segmented wormy bodies at a constant speed.

On the other hand, the luna had laid her eggs in the box within a known period from the occasion when I enclosed her in it until I felt satisfied with her parturient performance and gave her her release. That was roughly two days' time. If the eggs hatched over a more prolonged interval, say three days or more, I would be able to chalk up the point I had in mind. This would have been a simple matter for me to look into, but I did not do it. If I were to hear roars of "Why not?" I would give the same answer as I now proffer under calmer circumstances. I am not trying to make it easy for myself to say, for instance, that I had too much else to do, tending other kinds of caterpillar in other nets and prosecuting a bit of birdbanding in Stone Harbor. It is simply that it would not be good for the luna caterpillars were I to look at them too often, or—more accurately—to look so closely that I must disturb them. To peer in through the netting without touching it is OK, but you can't see very much that way when the caterpillars are very young and the uneaten leaves are still luxuriantly thick. Yet to time the hatching of eggs I would have to *remove* the net several times a day. This might dislodge some babies and I would lose them. Others that had just begun to take their first bites out of leaves might be frightened into immobility, for quiescence is the essence of avoiding detection by enemies. I long ago learned that the more you disturb caterpillars, the smaller (lighter) cocoons you eventually get. Therefore the *less* I molested them, the closer to normal my observations and results would be.

Well, then, I had to be satisfied with the notation that eggs were seen hatching on June 28 and 29. The shortest possible incubation period had been ten days—at least I could say that much. And one baby, which obligingly made its way outside the net through its meshes, disclosed that a luna enters this world as a minute green worm that is distinguished chiefly by possession of a horizontal black bar across each side of its "face." While the tiny larvae went about their ways which *may* have been as individually unique as those of you and me and two hundred other people, there was one item that I could pry into. The excellent glue that I have extolled had discharged its purpose perfectly and I was now able to examine the cardboard cutouts to learn how each egg had fared. Baby caterpillars do actually chew their way out of the shells, apparently eating the material, though I doubt that the parchment-like substance has any nutritive value. Most of them stop when they have eroded a hole big enough to let them out, though a few continue to eat and may consume so much of the shell that only a small disc in its droplet of dried glue remains.

Owing to said superior quality of said glue, I was enabled to complete a retrospective natal census of the 207 eggs originally tied on their carriers to branches within the net. All but 5 had hatched successfully. Only 1 of these 5 appeared to be infertile, containing liquid material that gave no evidence of having begun embryonic development. That is a very respectable rate of fertility and shows that the arrangement of the seminal receptacle in the female must be an efficient one. A student might nevertheless wonder why fertilization takes place that way: Why isn't the semen distributed to all eggs immediately at the time of mating? The simple answer to that one is that not all the eggs are ready on the female's wedding night. Perhaps half of them are, but if she is lucky enough to live out her entire potential life span of two weeks, she continues to ripen eggs (though at a rapidly decreasing rate) throughout all but the last few days of that period, and sperms must therefore be kept in storage for use when their availability is called upon on later nights.

I made a count of this process with several different female lunas—actually some of the siblings on which I shall shortly report in more detail. One female I chose to study from this standpoint (Number 80) was an especially large one. I was interested to learn how many eggs she could lay, what the fate of each egg would be and, since she had mated on her first night, whether the rate of fertility of her final eggs was any less than that of her first ones. If there were a decline, such a result would indicate either that her sperm bank had run out of funds or that the sperms had decreased in viability—and both of those eventualities could be checked by microscopic studies if anyone were sufficiently interested to make the investigation.

First let me go back and say that the four remaining unhatched eggs of my matriarchal luna had failed to be born into caterpillars at least partly because of me. The moth, confined to a box, had laid some of her eggs in dense clusters, and these four were so situated that although the larvae had gnawed holes in them, their exits abutted against neighboring eggs and the babies died of exhaustion in trying to overcome a second blockage. In the wild state, lunas lay eggs in short rows that would seldom present such obstacles to emergence.

My original luna's eggs, then, showed evidence of three ultimate fates: hatching, blockage, and infertility. Now, in doing postmortems on the eggs of her daughters, I had to devise further terminology to describe what had happened, and here I could actually attribute some of the findings to sibling variation. An infertile egg cannot be blamed on the embryo that never formed; nor can a blocked larva be held responsible for the awkward placement of its egg. But an embryo that begins to develop and then dies is certainly different from one that does not die. And an unblocked larva that succumbs in the act of eating its way into the world is also assuredly different from its more vigorous brethren.

As everyone knows, the factors causing variability may be both genetic and environmental and never will you completely disentangle one from the other. I considered the possibility that the extremely heavy rains we had in August might have drowned many eggs, causing their deaths at various stages after development had commenced, but is not genetics involved here also? If soaking killed some of them, why didn't they all expire?

I looked at all those eggs through my dissecting binocular microscope, so that as I cut them open with fine scissors (the unhatched ones, that is), I could easily tell how far each had progressed before something brought its further unfolding to a halt. Eggs that I took to be infertile were not only devoid of any formed contents but were filled with a characteristic turbid bluish gray fluid that was not seen in any of the others. "Live" eggs that had died early were still largely fluid inside, but on teasing out the contents I could eventually come upon a disintegrating shred that nevertheless showed distinct segmentation. Older embryos were naturally easy to spot at once in their juicy medium, while the ones that had completed development, whether blocked or "spontaneously" dead, reposed within dry shards. Luna Number 80's eggs had had the following experiences:

				Failed to Hatch				
Dates	Laid	Hatched	Infer-tile	Partly Devel. Died	Fully Devel. Died	Blocked at Emergence	Other	"Lost"
18th	313	282	4	0	4	19	0	4
19th	136	125	1	1	2	3	0	4
20th	85	83	0	0	0	1	0	1
21st	57	48	2	0	1	2	0	4
22nd	33	29	4	0	0	0	0	0
23rd	22	14	2	0	6	0	0	0
24th	8	6	2	0	0	0	0	0
25th	9	8	1	0	0	0	0	0
26th	8	7	1	0	0	0	0	0
27th	5	0	2	0	0	0	3	0
28th	6	0	1	0	0	0	0	5
29th	3	0	1	0	0	0	2	0
Total	685	602	21	1	13	25	5	18

Blocking at emergence was more pronounced in earlier egg batches because of massing of eggs at that time. Eighteen eggs that were "lost" simply disappeared from the net. That did not make any serious difference except for the five absentees on August 28, when the infertility rate was really becoming interesting. But I can say this: Sperms were probably still available and infertility now resulted from defective eggs. The three August 27 eggs and two on August 29 in the column headed "Other" were air-filled flimsy white objects, either containing a small aggregate of black material or being entirely empty. Not even the most robust of sperms could have hatched caterpillars from those. It thus seemed that fertility continued to be no problem and that sibling performances must be invoked to raise interesting issues.

That can be done if I continue slightly out of step in this tale, now mentioning luna Number 19's eggs, just to show that while results from one moth to the next may be somewhat the same, they are by no means identical. Numbers 19 and 80 provided me with eggs at about the same time, so the subsequent exposure of those living spheres in nets took place under approximately the same weather conditions. Yet the score for Number 19 was as follows:

Hatched	117
Infertile	4
Partly developed, died	46
Fully developed, died	6
Blocked at emergence	4
Total	177

One must immediately question why a quarter of Number 19's eggs should have died after beginning growth. In human embryology, one of the commonest causes of *early* abortion is faulty development, one or other of the vital organ systems failing to establish itself properly. Perhaps the same can happen with lunas. As a matter of fact, it may be more than coincidence that the proportion of dying eggs was 26 percent here, for that is exactly what one would expect if a so-called lethal gene had come into play with the chance coition of the pair of moths producing this batch of eggs. In any event, something caused more of Number 19's eggs than of Number 80's to give up their little ghosts, and that in itself was a manifestation of variation between the two sets of parents. (Let me anticipate only once more by saying that both fathers and both mothers were siblings—or, according to the remote reservation I have had to make, they were at least half-siblings.)

Now I must return to the original netful of larvae which, on July 12, could be considered two weeks old and might really begin to reveal individualities. In fact they did—or *one* did! There had been a shocking postnatal mortality, from causes which I failed to determine. Out of 202 eggs which hatched, I could find only 150 caterpillars. The count on this date was coincidental with defoliation of the persimmon branch, and while transferring the net to a new position I naturally made a meticulous census. I did not move caterpillars individually or even touch any of them. It was sufficient to clip leaves or twigs, where the larvae clung, with pruning shears and drop such snippets into the net.

They would soon leave their former perches, especially when leaves began to wilt, and climb up the sides of the net to new positions on fresh leaves.

In making that historic move and scrutinizing every caterpillar to see whether it would "say" anything to me, I finally encountered one which, even at that young stage, was *very small* compared to the others. Great! At last I had a concrete example of variation! I wryly thought that 52 larvae, mysteriously dead in their babyhood, were variants of a sort, too, but I could not work with defunct material. Here, in this runt of the brood, lay the possible beginnings of experimental adventure—if only it were not the last survivor of sickly variants and itself about to join the others in their premature extinction.

Within the confines of a net, which was really a vast labyrinthine space to a tiny larva confronted with a maze of walkways along twigs that bore a multitude of edible leaves, I was unable to keep track of the "tiny" one, and although I eventually harvested some very small cocoons (of which I shall soon speak), I do not *know* that the early recognized runt was responsible for and dormant within one of them. It *may* have died instead, and further variation among surviving siblings might have given rise to later divergences in size that caused a new larva to take up a position of last place in the physical culture contest. Or the original runt might have caught up and eventually become indistinguishable from his average companions.

Mortality continued to rob me of caterpillars as they maintained their growing-up. Every time I had to change the net to a new branch—an increasingly frequent operation as consumption of leaves gained in rapidity—my tally became lower. Sometimes I now could discover what had happened to a larva, for larger corpses became more conspicuous than tiny ones had been. Occasionally predacious bugs sucked caterpillars dry. Even though I tried to exclude such predators from the net, they could insert their long beaks from the outside and deflate a caterpillar that happened unfortunately to be within reach. Sometimes a loss was my fault. Once I failed to close a net snugly at its neck and several caterpillars escaped; another time I closed it snugly enough, right on a caterpillar that I had not noticed, and when I changed the net next time I found the victim squashed where I had slain it.

To make things a bit easier, I divided the larvae into two lots and put them in separate but adjacent nets. This should have saved me trouble, and of course it really must have, but I was scarcely aware of it because of explosively increasing appetites: I still had to change the nets as often as before. Now I was becoming mesmerized by the occupation—resigned, conditioned, or otherwise fixed in the notion that life would henceforth be one continuous transfer of luna caterpillars from a denuded persimmon branch to a fresh one, with attendant care of torn nets and entry of appropriate diary notes on my work sheets.

Censuses ran from the 150 larvae I have mentioned on July 12 to 144 on the sixteenth, 140 on the twentieth and 139 on the twenty-second. Then the next inevitable step, whose possibility had almost been forgotten in my absorption with caterpillar alimentation, restored me to a less phantasmagoric state. On July 24 the first 18 larvae spun, or began to spin, cocoons.

At this important juncture I was down to 128 caterpillars. Moreover, I found it difficult to sort out truly small from large ones, for those that looked small were still feeding and might become as hefty as their larger siblings by the time they, too, spun cocoons. Therefore such judgments must be reserved until all feeding had stopped and the cocoons could be weighed. Meanwhile I was nevertheless able to make an exceedingly welcome observation. A mature luna caterpillar is a quite beautiful object, plump and green with a narrow yellowish stripe down each side. At least that is what most mature luna caterpillars look like. But among my 128 larvae, two were not green but, rather, of a reddish color that I called wine. Number 19 was especially rich in this color, while Number 80 exhibited it very plainly but somewhat less intensely. These are the two lady lunas I have already mentioned with respect to their eggs. At the time of their spinning I did not know, of course, that they both would be females (though their cocoon weights predicted it). If they were of opposite sexes, and if they emerged at the same time, it would be inevitable that I should try to mate them in an effort to establish a strain of wine-colored larval lunas. Since I could not do that, I did mate them each with a brother in hopes that something of interest might appear in the progeny.

From now on I could follow every luna organism as an individual. I had had Rudd make a wooden cocoon box, divided into numbered compartments and provided with a screened top. As cocoons were spun in the nets, I harvested them day by day, weighed each one, gave it a number in the series and housed it in the appropriate compartment of the box. Actually I did not know the spinning dates of every one, but the information was adequate because I knew not only when the first caterpillars had enshrouded themselves but also the last date—July 29—when a single remaining larva could still be seen munching on a leaf. Thus the spread of time over which spinning began could be recorded as five days, and a spread of three days had been added to the original two-day interval separating the oldest and youngest eggs. This means that the shortest possible larval period might have been twenty-five days and the longest thirty-two, though the lower limit was more likely twenty-six.

As the last statement shows, I was now confronted with some variations that I could not positively define but had to cite only as estimates or approximations. Very well, they were variations just the same, for it was obvious that the caterpillars had had different histories and were now occupying recognizably individual positions in my tabulations. Next spring, when they emerged as moths, they ought to display even greater disparities.

Next spring! My little *Golden Nature Guide* mentioned that lunas were double-brooded in parts of their range but did not say where. That left-front wing in the orchard on August 12 last year should have put me on the alert, especially because—as I now reexamined it—it did not have the purple outer margin characteristic of the spring brood. My matriarchal female this June had been purple-bordered, but her first son, who appeared in Compartment No. 14 on August 9—almost exactly a year after "the wing"—obedient to the book, bore a handsome yellow rim on his front pennants.

Now I was in business all over again, and really with a vengeance, because next year's experiments must suddenly be done now when I was in the midst of rearing Polyphemus, Cecropia, and Io caterpillars as well. I had Mrs. Fisher, my cleaning lady, sew up a bunch of new nets and prepared myself for full abandonment to the pleasures and suffering of luna cultivation.

The ensuing two weeks provided me with as much variation as I ever could have hoped for. To take an example, the five-day "spread" separating the first and last spinners was now extended to a twelve-day interval between the first male to emerge on August 9 and the final three females that stepped forth on their purple legs on the twentieth. The actual duration of pupation varied between sixteen and twenty-one days in the cases of seventy-three lunas whose spinning dates had been recorded, but those individuals did not include the final laggard larvae that put off cocoon-making until the very last—endomorphs, without a doubt, who had no stomach for a future life of permanent dietary abstinence.

Already I began to be confused by the numbers of variations that were disclosing themselves, for there were others that I have not yet had time to mention. How could I deal with them all, without becoming buried under mountains of detail that ultimately had little to do directly with lunas but only sounded as if the moths had taught me something profound? One could, for instance, reduce everything to averages, viz., the average female weighed so-and-so much; the average male emerged on August 14. Apart from thereby cancelling out the variations I was seeking, I would also arrive at statistics that stood for nonexistent creatures. If you know that only one person in ten gets hay fever, you can not persuade a hay-fever sufferer that he has not got it because the average person doesn't.

My lunas were similarly un-average, individually and vibrantly unique in their qualities. To be sure, averages *did* help, if only to show me by what degree each moth departed from them, in either the positive or the negative direction. I resolved the threat of total confusion by first dealing with a single trait at a time and then seeing how it could be compared with other traits. Thus some characteristics were found to be generally related, e.g., a light cocoon was likely to yield a male rather than a female, and a heavy moth usually had a wider wingspread than a somewhat less hefty one. But the making of rules was definitely out of order, because individual exceptions turned up all the time.

The traits presenting themselves for enumeration were of course endless. Since I did not have endless time, I selected several of the more obvious ones. These were: sex, weight, duration of pupation, order of emergence, and wingspan. Even some of these were not easy to define in the first place or else had to be accepted on the basis of an arbitrary definition. Let me begin by asking you how you would weigh a luna moth. To start out, you would almost certainly have to handle it to get it into a suitable position or container for being weighed, and that would mean disturbing it. Now disturbing newly emerged lunas—or any kinds of moth—is something you *don't* want to do. If the wings are still soft, they will crumple and then inevitably set in an imperfect shape. If enough time has passed for the wings to have become hard, your interference will set the moth into motion and it will try to escape. Apart from such frenzied activity's playing hob with the pointer on the scales, you may not be ready to release the moth—you might want it to mate or to lay eggs—so from all standpoints you would rather calculate the moth's weight in an indirect manner.

I did this by weighing the cocoons soon after they were spun. The figure I recorded therefore included something besides living contents, for the silk weighed something appreciable, even in such a flimsy cocoon as luna's. But in addition the caterpillar, having shut itself into wrappings for protection during further metamorphosis, shed its skin in becoming a pupa, and later the moth left its pupal shell behind when emerging. The moth also ejected a bit of waste fluid into the cocoon, but I have no idea how much that weighed, for I did not put the empty cocoons on the scales until they had thoroughly dried out. Therefore, with the admission that I was off by a drop or two of fluid, my "corrected" weight for a moth was the fresh cocoon weight *less* that of its dried cocoon containing two former dresses of the flying creature. And *that* failed to allow for loss of substance attributable simply to being alive from moment to moment: carbon dioxide and water liberated into the atmosphere did, after all, subtract a mite daily from each slumbering "doll."

Similarly, I simply adopted a wing measurement which I accepted as representing the flat dimensions of the moth, as we distinguish between the height and weight of human beings without bothering about all the other solid and linear measurements that might be taken (except in beauty contests). This was purely a matter of convenience, for when a moth finished with whatever chores I had given it, I did no harm by picking it up at the last minute for a quick check with a ruler before letting it go. For this I used a transparent plastic millimeter stick that I could overlay on a front wing and quickly read through it from the wingtip to the thoracic junction. I called this measurement the costal chord, the front margin of an insect's wing being known as the costa, because I was recording the straight-line distance, not its curved edge.

The problem now—or perhaps I should call it more recreation or delectation than a problem—was to see how the various variable traits were related to each other, if at all. One correlation which did *not* become established first jolted me and then struck reflective second thoughts as demonstrating individualness as perfectly as anything could have done. I had approached the analysis of pupation periods with the prejudiced view that rapid metamorphosis, commencing in these summer cocoons immediately upon transformation of a caterpillar into a pupa, would culminate in the formation of emergent moths on a smooth time scale in relation to weight. In other words, I presumed that the lightest cocoons would transform into moths most quickly simply because they had the least amount of "caterpillar material" that had to be changed over into "moth material." After all, that process involved cell divisions and organic chemical reactions (or metabolism, if you will), all of which could probably be reduced to quantitative formulae in relation to the work to be done at circumambient temperatures and in accord with thermodynamic laws. Surely here was one arena in which variability would be denied to the lunas, since their vital substances in this instance must respond to physico-chemical rules on an inexorable milligram for milligram basis.

But not at all! When I tabulated the corrected weights of forty-two male and thirty-one female lunas whose pupation peri-

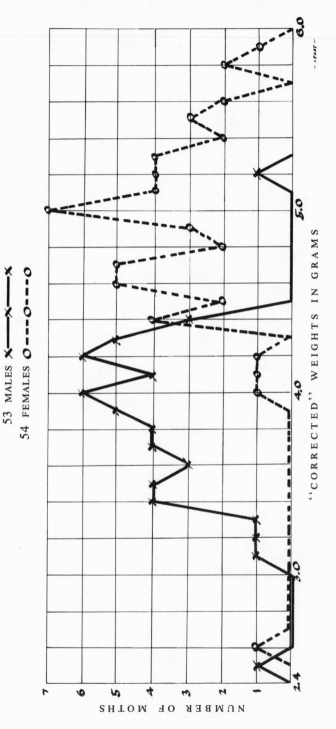

"CORRECTED" WEIGHTS OF LUNA MOTH SIBLINGS

53 MALES ✕——✕
54 FEMALES ⭕----⭕----⭕

"CORRECTED" WEIGHTS IN GRAMS

NUMBER OF MOTHS

ods were accurately known, I discovered that there was virtually no connection between the two, thus:

Days of pupation	Number of males	Average weight of males (in grams)	Number of females	Average weight of females (in grams)
16	1	3.4	1	4.5
17	10	3.9	4	4.8
18	11	3.8	0	
19	10	3.9	10	4.7
20	7	3.4	11	4.9
21	3	4.1	5	5.4

And of course, as I have already said, averages are sure to cloud or to lose more direct observations that involve the performance of exceptional individuals. For even one moth that developed very rapidly or very slowly, compared with the average, would disprove all the rules of chemistry and physics that proclaimed what it should have done instead.

I had to conclude that pupation periods varying from sixteen to twenty-one days in cocoons of roughly uniform weight were true indications of genetic variability. Metabolic processes were not quite as immutable in their rates as I had supposed but were, rather, under at least partial control by whatever centers in lunas direct such biological processes. Thus, if I had been able to begin with a set of eggs that had all been fertilized simultaneously and had thenceforth remained in step throughout caterpillarhood, I would still have encountered emerging moths in the cocoon box on successive days for almost a week.

One can detect what appears to be a difference in metabolic rates between males and females. In the foregoing list, for instance, the ten males that took nineteen days to metamorphose into adults weighed eight-tenths of a gram less, on the average, than ten females that took the same amount of time. It would thus appear that males are slower, or females faster, per unit of weight (according to whose side you take). I suspect, however, that this difference is misleading, for whereas a male emerges with (so far as I know) all his body material fully transformed, females have matured only a part of their eggs and continue to develop the immature ones for days afterward. Thus at emergence females may have converted an amount of material that is after all no more than the equivalent of that which males have metabolized.

Well, then, let's look at some other features that *do* conform to some sort of order that appears to make sense to us. Mind you, we can't impose our notions of "sense" on lunas, for creatures of every sort have the right to live lives that appear insane in human terms. But on the other hand, nature is basically orderly, and it may be only trivialities that violate what we expect and demand of amoebas and anteaters. Luna's tailed hind wings, to be sure, are utter nonsense when it comes to flying, but other aspects of the beast are humdrum enough.

Emergence dates, for example, tell a more interesting story than I have revealed in referring only to pupal periods. Though the latter differed from each other by a maximum of six days, variations in the time of cocoon-spinning spread out actual emergence dates to twice that long, so that I found fresh recruits in the cocoon box over a twelve-day period. A tabulation of these individuals by sex, regardless of other facets of their histories, showed a striking predominance of males during the first few days. Thus, between August 9 and 13, twenty-eight males but only six females made their debuts. It was not until the fourteenth, which incidentally was the peak day of emergence, that females threatened to predominate with a tally of fifteen against eighteen of their brothers. From then on the ladies had it their way, though one could now have predicted that result if the sex distribution was destined to be equal (as it finally was: sixty-two males to sixty-four females).

But wasn't there something vaguely familiar about those results? Of course: It all harked back to my initial experience with these siblings' antecedents of a generation before in the bird nets. From June 5 to 17 I had caught ten of their uncles but nary an aunt and it was not until after this that their mother had turned up.

With two examples of the "average" earlier emergence of male than female lunas, in both the spring and the summer broods (and this is true for other saturniid moths, too), one can safely

infer that male caterpillars are in general the earlier spinners.
One can not discern the sex of luna larvae externally, but on the
two bases of first spinning dates and light weights one could ordinarily select male cocoons with a fair proportion of accuracy
from the total in a given batch. One may again question whether
this is merely a physiological happen-so or whether it serves
usefully in luna's economy. It surely means that when females
emerge, males will already be on hand to accommodate them,
and one *can* discern an advantage in that arrangement. After a
male's duties have been performed, he may as well be dead as
fluttering around lights, so far as the future of the species is concerned. But females remain useful for their entire lives as long
as they continue to ripen eggs (even though on the last days
these are few indeed). Hence, since females' careers are limited
by their inability to eat, it is important that they should mate
on their *first* night, while it does not matter whether their husbands emerged on that same day or a week ago, provided they
can still manage to reach spouses and fertilize them with a final
expenditure of depleted energy.

So far as actual weights, apart from any other consideration,
are concerned, lunas show exactly what one would predict on the
basis of females' being loaded with eggs at emergence: Feminine
members of the race, on the average, exceed their future husbands in heftiness during the virginal state. (There is but little
irony in the observation that they rapidly slim down as they lay
eggs, for they stimulate no further interest in men and are not
much longer for this world anyhow.) In the diagram I have composed to show how these weights are distributed, it is easy to
see that there is a definite zone of overlap between four and
four-and-a-half grams, where the heaviest males and lightest
females weigh the same. Also one can discern a peak average
weight for each sex, at slightly under four grams for males and
about five grams for females (3.9 and 4.9 grams respectively by
tedious but exact calculation).

One might at once conclude that the difference of one gram
must be equivalent to the average egg load of each female. I wish
that I could confirm that suspicion by reporting that I had
weighed a series of females first at emergence and again at their

points of expiry, but to tell the truth the notion of doing so did not occur to me until much later when I began juggling figures to see what my blindly taken notes might reveal. Thus, all I can say is that a female Luna of ordinary size does carry at least one gram of eggs. But it is possible—and even very likely—that the numbers of eggs eventually laid greatly exceed one gram in weight and that the slimming effect of oviposition brings the matrons' bulk down to levels below the standard set by males. I have a really tricky device to support that suggestion.

They say that one should not apologize for one's deficiencies. I suppose that is because such lacunae might go unnoticed if they remained unmentioned. However, it will be so obvious that I am not an aerodynamic engineer that I *do* apologize, if only to make myself feel better while making the following comments. I think it is safe enough to make the rudimentary assertion that wing size is related to the load to be lifted. That is true at least if we stick to the same general design of aircraft. Therefore a heavier luna ought to have bigger wings than a lesser moth. By "bigger wings" the engineer would refer to their area, and he would speak of the load they could support per unit of wing surface. Now it is far from easy to measure the area of an object as peculiar in shape as an insect's wing, and I was certainly not prepared to do so. In selecting the costal chord as an index of size, I made the assumption that this single linear dimension would, in fact, supply a valid notion of a wing's aerodynamic properties, because shape remained constant even while size varied. In proceeding with my analysis, therefore, I shall be totally in error if that assumption is false.

Nevertheless I have drawn a pretty picture (see page 97) to show how the costal chords line up with the weights of male and female moths. To be sure, heavier lunas of both sexes tended to have bigger wings than their smaller siblings. But the average wingspans of males and females, as shown by broken extended lines in the graph, fell on different, though parallel, gradients, the trend being for females to be smaller, considering their bulk, than their brothers. To me this indicates that a female, on completion of mating, must set out on her initial egg-laying flight with considerable difficulty, flapping her wings laboriously and probably quite rapidly in order to remain airborne. As she discharges eggs here and there, her buoyancy increases until, at a certain point, she begins to sail "with the greatest of ease," just like her erstwhile husband. At this time a group of females would fit the picture of costal chords for males in the diagram and its extension. But let them lay additional eggs, and their aerodynamic formulae would be carried to a new realm of ethereal fluttering, unknown to males, and such as they never dreamed of either on the wedding night. Thus, though the females become ever more feeble, their task is lightened for them at each brief stop at a suitable leaf, and the joys of ovipositional flight might go on to the vanishing point if one could only transform one's entire substance into eggs.

My luna-moth siblings were more than inert gemstones to be weighed, measured, and admired. They were, after all, very much alive—I could even say acutely alive, because such destinies as lay before them were intensely immediate ones whose delay in fulfillment spelled only waste without hope of restitution. My tampering with their persons and their affairs was therefore scarcely more than incidental to their purposes. At any rate, I did my best to intrude as little as possible and when I did so force myself upon the lunas I tried to make my manipulations brief and even useful.

"How could I ever be useful to them?" one might easily inquire. In truth I had already served them outstandingly well, in spite of the artificial existence they had shared since their mother laid eggs in a box and their caterpillars grew up in a net. Because of me the matriarchal luna was represented in the next generation by 126 adult progeny (two of the cocoons refusing to hatch), and that was possibly 125 more than would have survived in the wild, exposed state, had the mother herself lived long enough to lay that number of additional eggs if I had simply liberated her from the bird net.

And now I hoped to serve the siblings further by continuing to protect them during their mating period. Here 1 was up against another unknown in the lives of lunas, due to my not having reared them before. What I hoped to do was the simplest

AVERAGE WEIGHTS AND WING LENGTHS OF LUNA MOTH SIBLINGS

45 MALES ○—○—○

46 FEMALES ●

HYPOTHETICAL EXTENSIONS ▬ ▬ ▬

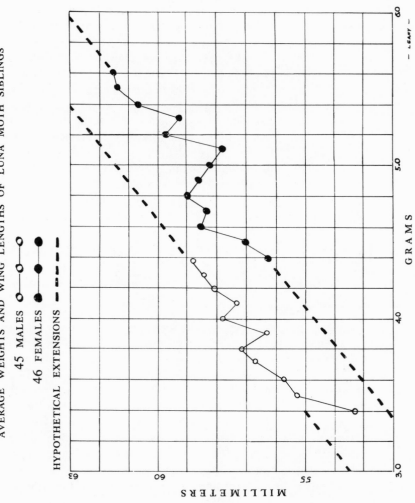

of all possible acts under the given conditions. In the evenings, after a day's hatch had emerged and hardened its wings, I would tally the occupants of various compartments in the cocoon box and then write out a mating roster: Female 35 x Male 16; Female 47 x Male 54; and so on. I would lift the wire-mesh lid and quickly transfer males from their hatching compartments to those of the appropriate females. After that it was up to luna nature to take its course and, since most of the other saturniid moths of my acquaintance usually prolonged their midnight mating into afternoon of the next day, I hoped to be able to peer into the box before breakfast on the following mornings and behold the various couples still joined in copulation.

Some organisms quite understandably refuse to mate if they feel that they are confined. Others seem actually to require a large amount of space in which to maneuver before the mating impulse can be released in them. These creatures are said to be eurygamic, in contrast to stenogamic ones that will mate in a matchbox. A virginal queen bee must be the classical example of a eurygamic personality since she disdains drones in the hive but will accept them only while both are in flight. If lunas were eurygamic, they would surely not recognize a cubicle of the cocoon box, measuring only about four by four inches square and five-and-a-half inches deep, as a honeymoon suite.

However, I had some reason to hope that the species might be stenogamic, at least so far as the females' attitude was concerned. One of my boyhood books was called something like *Butterflies and Moths.* That was not it, but I do know that it was written by Ellen Robertson Miller. In her chapter on lunas, she spoke of tying a virginal female overnight by a string around its waist to a branch in her cherry tree, and of a male's being present the next morning. Surely a tether of that sort must have made the female feel very stenogamic; yet she had not been too upset to follow her natural impulse to secrete the volatile chemical mating lure that had drawn a mate out of the circumambient darkness. That seemed to take care of females, but now I was confining both sexes, and it remained quite possible that that would alter the entire story. For insects are extraordinarily automatic and nonadaptable creatures. A slight alteration in the way they do things is often enough not simply to throw their entire behavior patterns out of balance but to bring them to a total stop.

Obviously I was very keen on these matings' taking place, for the matching of known brothers and sisters might intensify some of the genetic characteristics innate within this stock. If I now had to tie out my females on string tethers, who knows what wild riffraff might come to mate with them? And then there was no telling what new traits would be introduced. I would be back where I was in the beginning, dealing with no more than first-generation siblings which, though absorbing enough, would never take me deeper into problems of luna genetics.

I need not have worried too much on the stenogamic score —at least I think I need not have. The first number of matings I arranged came off perfectly, though later ones were successful in only half the cases, so far as I could tell. For reasons that will soon appear, some of the unsuccessful unions may actually have taken place but they must have been brief and, contrary to conventional rules, the moths may have separated before daybreak, in which case I could not tell whether they had mated or not. But at first everything seemed to be very fine indeed, and I took much pleasure in releasing yesterday's well-fertilized females in the late afternoon while arranging a new batch of nuptials for the coming evening.

Of course I saved some eggs—I shall come to that in a moment. However, I was swamped in riches. It would be impossible to study the progeny of more than fifty pairs of sibling lunas unless I had an army of employees to assist me. Sewing up caterpillar bags alone for that number of larvae would occupy all of Mrs. Fisher's time, in which case no dishes would be washed, the house would remain untidied, and the cost of all that netting would run me bankrupt. Besides, I did not have enough persimmon trees and I preferred not to switch to sweet gum, hickory, or walnut because that would introduce an unwanted variable in my experimental setup.

Thus the liberation of most of the females was an act of joy that I felt both they and I had earned. On reflection I realized that in a way this was wrong: Ordinarily, without my intervention, there would not have been so many lunas on my farm. I was now changing the natural balance of things. Yet nature, in the guise of a male Cardinal, was quick to recognize the surplus of lunas that appeared daily near the cocoon box. Soon the bird arrived as regularly in the afternoons as if I always set out sunflower seeds at that same late hour every day. The lawn became littered with wings, some of them *left front* ones, bringing an outstanding cycle of my experience to a most convincing close.

The moths having tricked me in the first place by emerging in August instead of waiting until next June, and now swamping me with invitations to further study, I could afford to be selective in which ones I should use as the new progenitors of captive stock. For it was unthinkable to drop a project that had not only begun with great fanfare but promised to prosper twice as fast as I had expected with its two yearly generations rather than only one. As I have already said, the two wine-colored caterpillars had immediately suggested a trait for me to select, to see whether it had a genetic background. When they both yielded females, I was not too greatly disappointed because even if they had been of opposite sexes, a further chance would have had to operate successfully to give me the opportunity to "play" with them: They would have had to emerge at the mutually best times for their mating, and since it was unlikely that all these conditions would have been met, I had really not thought seriously about the cross-breeding possibility. Instead I was prepared to be quite happy if both females consented to mate with their brothers. Since they did, I blithely undertook the task of rearing their offspring, both to test out the wine-color trait and to assure myself the possession of cocoons next spring when, as a result of brother-sister mating or perhaps simply from further chance for variability to display itself, I might be able to refine the strains already established or choose new ones to exploit.

But before becoming too deeply immersed in the wine trait, I must mention dwarf. Everything I have just said about the unlikelihood of two wine-colored larvae yielding a male and a female at the proper time must now be unsaid with regard to another couple which performed those improbable miracles. I did not spot them as caterpillars, because what might look like small ones could easily have been merely slow feeders (as already suggested) that would in the end catch up to their siblings as late spinners. It was only after I had harvested the cocoons and was routinely weighing them that these two asserted themselves as exceptions in the series. You can see them represented at the far left of the graph (see page 91) showing corrected weights of the males and females. These are way off by themselves, and at the time I discovered them on my scales, I vaguely thought two thoughts: "They must both be males," and "Possibly they are feeble and neither will hatch."

Naturally I could not then be certain of their sex. It would have been possible to cut the cocoons open and examine outlines of future antennae on the pupal shells. That would have given me positive information, but I have found that such interference or intrusion may reduce the chances of successful emergence later on. Since I had no premonition of what was going to happen, I restrained my itchy curiosity—otherwise I might not have been able to.

The future female was cocoon Number 99. I marked it with an arrow on my work sheet at the time of weighing, to keep my attention directed to the fact that it was so light and would yield an especially small moth if it ever hatched at all. The other, Number 92, I marked with an arrow also, but made the additional note that the cocoon was dark and dirty. Thus I was clearly not looking at these objects with much hope. They impressed me as being bizarre rather than auspicious. The best to be said for them was that when it came to variation among siblings, they provided splendid extreme limits.

I can only understate my feelings by averring that I was amazed and awestruck on August 17 when both cocoons hatched, one of them yielding a female. Apart from their diminutive proportions there was nothing unusual about either of them: Each was a perfect little specimen. The male, of course, was not greatly out of line with his next-bigger brothers: his costal-chord measurement of 50 millimeters was actually only a single millimeter less than the nearest rival of his sex for the runt size prize. Still he *was* the champion dwarf and therefore an ideal mate for the female. She, however, was far, far below minimum standards in her department. Her costal-chord measurement of 48 millimeters was even less than her intended husband's, while her most petite sister flapped an ungainly 54.

The big question now was whether these two really were normal in other respects, particularly as regards their abilities to perform conjugally. If they should go through the motions of mating successfully, that was still no guarantee that subsequent eggs would be fertile or, if so, that they would develop without mishap to the point of hatching. And after that there would continue to be questions of whether these moths were dwarves on the basis of heredity or only of environment. If their caterpillars had remained small simply because some force (of which I was ignorant) had prevented them from feeding properly, their progeny now might stuff themselves to regulation size or even result in some overweight individuals.

Actually the progeny of Number 80 of the wine strain would give me a good check against descendants of the dwarves, because that moth had been extra large and I had been able to mate her with an extra-large male. Thus, while they were serving in investigation of the wine trait, they could be considered also as candidates for appellation as giants, though they were not really uniquely outsized. The difference between the two pairs can be appreciated when I mention that Wine Female 80 and her husband-brother both had costal chords of 61 millimeters, while Wine Female 80, at a corrected weight of 5.8 grams, weighed more than the combined mass of the dwarves, which was only 5.1 grams!

Those divergent standards would be handy indeed. Imagine the most perfect results that could emerge from the setup I was now contemplating. In two nets on adjacent persimmon trees I would rear, in the first instance a race of gigantic wine-colored caterpillars, and in the other a strain of tiny green ones. The worst result (though it would not demoralize the world): two nets with caterpillars of identical color and size. Then there were two intermediate possibilities, both of them more reasonable and perhaps more likely. The one that I felt I might hope for without being completely wild would present itself as a visually obvious distinction between the two batches of caterpillars—a fair sprinkling of wine-colored ones in a net containing largish larvae, compared with only one or two departures from green in another net containing smallish larvae. Anybody could look at such a result and recognize it as a real one. What remained was the statistically positive result that could not be appreciated until a computer had manipulated data such as the combined cocoon weights of the two series. In that case the figures for dwarf and giant would overlap to a considerable degree, and one would wonder whether the extremes could still be considered as the result of genetic crossbreeding or whether all the specimens formed a continuous series on the basis of natural physical variations induced by social interactions in the net. Tests for statistical significance that come out positive determine the course of action of many schemes, but I loathe the experiment that has to be read that way.

But here I was dreaming of dwarf progeny before I had even put the little moths together! Talk about schemes of mice and men! I looked into cubicle Number 99 on the morning of August 18 fully confident that I would behold a blissful joining of the two. After so much luck on the constructive side, I simply had the feeling that things would continue in that atmosphere or trend. They must be so ordained in the stars, or else some obstacle would have broken the charm before this. Yet it is possible to incur late failures, too, as I now rudely realized. With real shock—the greater because I had run out to the box before stoking up on breakfast—I saw the lunas perched on opposite walls of the cubicle. Worse: They were both battered, as if they had spent their night trying to escape rather than in loving cultivation of each other's company. Worst: The female had laid a few worthless eggs on one side of the cubicle. Since she was small, she did not contain many eggs to begin with, and I hated her to have wasted any at all.

However, she might consent to mate tonight. I faced an important question: Should I try her with a different male? That would scale my experiment down to a considerably less interesting one, for the dwarf factor would be diluted by whatever greater wingspan an alternate husband must inevitably possess, despite the recommendation one could still give him as a brother. No: I would persist with the dwarf male, risking an all-or-none outcome of the mating attempt.

I hate to transcribe my notes of the following days, but it must be done. August 19: "The dwarf pair did not mate." August 20: "Dwarves did not mate." On August 21 I did not even bother to make a record. Meanwhile the female continued to lay some eggs every night. On August 22 I gave up, since both moths were beginning to become feeble. The female had laid about a hundred eggs, and her abdomen (which I now dissected) contained only about 15 partially developed ones—another measure of her small stature when you remember that Wine Female Number 80 had laid 685 eggs!

Simply as a gesture of stubborness, I scraped the eggs from the walls of the cubicle, gathered up detached ones lying on the bottom, and put them in a gallon jar with some fresh persimmon leaves. Now listen to my note for August 28: "By great fortune I looked into the jar this morning and found some of the dwarf eggs hatching!" (There is no need to copy the rest of my paeans.)

Puzzlement visited me, but it was neither deep nor prolonged. No one in the world would believe that a virgin birth had occurred when a male was in attendance on the dwarf female at all times. Therefore, if he did mate with her on the sly, the question is "when?" The shortest incubation periods for luna eggs in my records, following recognized matings, were around ten days. According to that timing, I now concluded that the dwarves had mated on their first night together. The only difference in their behavior, compared with their larger siblings, was that their union had been much briefer than normal—unless, of course, some other pairs that I had classified as eurygamic had in fact also mated but finished the act before my early morning inspection. But I doubt the latter possibility. Some of those females, suspected of not having mated, were kept in the cubicles for an extra night with different males, whereupon they, too, laid some eggs on the walls of their cells—eggs, that is, which never hatched.

No. My dwarves had simply mated quickly, and that brought up an interesting point. Perhaps small size was associated with a speedup of various functions, in the way that a mouse's heartbeat is faster than that of an elephant. If the moths produced dwarf progeny, I might be able to test the next generation for speed in respect not only to mating but perhaps in other features as well, beginning with duration of the caterpillar stage and going on into adult longevity and so on. But thus far I could point to two factors that were *not* different: Incubation of the eggs had extended to the standard length of time while the eggs themselves also looked to me normal in size. If the dwarf female had laid 600 dwarf eggs instead of 100 regular ones, that would have been a real departure! In that case I would truly have expected the eggs to yield nothing, for they would have then been monstrous, in the freak sense, and my moths were, after all, not that greatly removed from recognition as conventional members of their species.

Not that monsters aren't a possibility to be kept in mind, even when one is playing with nothing more than luna moths. On August 15 one of the emerging siblings, a male, provided me with material that would have looked respectable in any museum of pathology. Had I found this one's *right front* wing in my orchard a year ago, I might have been excused for announcing to the scientific world that my farm supported a veritable dwarf species of luna, for the costal chord of that right wing measured only 43 millimeters (as compared with 48 for the dwarf female and 62 for my largest sibling). But this wing was attached quite normally to the remainder of a quite normal moth whose left front wing measured 57 millimeters. One could easily see that the moth was built along the lines of the larger wing. Something had happened to cause the right wing to remain small. This force, however, appeared not to have been an injury because the wing was not in the least distorted or misshapen, being simply a perfectly formed but diminutive wing. I opened the cocoon to look at the pupal shell and found that the two sides were correspondingly different, so that the small wing was preordained at the time the caterpillar had cast its skin to assume the quiescent stage of transformation.

This looked to me like a mosaic, which is a peculiarly mixed genetic condition in insects wherein the parts of the mixture do not blend but express themselves independently or piecemeal in different areas of the same organism. When the mixture happens to involve sex, one can see bizarre combinations of maleness and femaleness, involving the right and left sides or the various appendages.

But never mind that. My dwarf caterpillars were hatching, and the leaves that I had put in the jar several days ago had partially dried. They were anything but suitable for tender jaws. I must quickly put the young larvae and the remaining eggs on a living branch in a net. That was done immediately, but apparently I carelessly included a few predacious stinkbugs in the net's interior, actually confining them so that if they became hungry they had no choice other than solid-gold luna caterpillars for lunch. By the time I suspected the presence of those predators a week had passed. The eggs, having been laid over several days in the cocoon box, hatched over a similar period of time, but whenever I looked into the net I was puzzled because I was unable to detect an increasing population of larvae on the leaves. I was sick in mind when I saw those fat pentatomid bugs, their green juice derived from dozens of tiny succulent baby caterpillars, and it gave me no satisfying surge of revenge whatsoever to squash them.

At one point I became so distressed that I spent a day pampering four dwarf larvae that would have been a loss otherwise. These had hatched from loose eggs, scraped from the walls of the cubicle, and in emerging from their shells they had come right through meshes of the net. Now on the outside of their intended sanctuary, they would either fall to the ground or else crawl to an exposed branch (if they could get that far) with all the attendant dangers of an open environment.

I brought these four strays into the house and placed them on a spray of fresh persimmon leaves in a glass of water on the dining room table. Instead of being happy that I had transferred them securely to a haven of safety, they did their best to escape. I had to go away at various time, and on one of my returns I found two of the caterpillars roaming at opposite ends of the table and a third missing. Only one was left on a persimmon leaf. I restored the two wanderers to the sprig and one of them finally settled down with the least adventurous one, and these two at last began to eat, which was a highly favorable sign. The other rover would not compose itself at all. I almost lost it many times, either when it fell into the water of the glass or again as it wandered across the table and off its edge. At nightfall that one finally settled down on the midrib on the upper surface of a leaf, and though it looked as if it tried to gnaw a couple of times, I could not see that it got anything to eat and it passed no frass, which the other two now were doing regularly.

Only next morning—almost twenty-four hours since I brought it in—the errant larva began to incise the margin of the leaf. I would have expected it to have become too weak by that time, or to have become desiccated or to have died of starvation. However, it seemed to have extraordinary stores of energy from its tiny egg, to have remained so active for so long, granted that it did spend long periods resting.

What did that incidental episode mean? The only thing I could think of was dispersal, but that did not make much sense. The moth, in laying her eggs, should have taken care of dispersal much more effectively than hatching larvae could. Also, if she had laid eggs on the proper kind of leaf, the larvae ought not venture to desert it. Maybe I was only reaping from what I had sown: an abnormal reaction following the artificial conditions I had provided. For the caterpillars at hatching had *not* found themselves on edible leaves. Discovering that the world was not one vast gingerbread house, perhaps they became sceptics and felt leery even of persimmon when that foliage first came underfoot.

All this happened before I had become aware of predatory bugs in the net. Thinking that I had now satisfactorily established three baby luna caterpillars, I carefully put them *inside* the airy bag where they belonged, and thus probably consigned them to quick death.

I now had three batches of late summer luna crawlers: Wine Number 19, Wine Number 80, and Dwarf. Whether it was the season or the fact that I cared so much more desperately in August than in June I can't say, but it seemed more difficult to rear these caterpillars than had been true with their antecedents. The weather certainly became less favorable, for it soon was Septem-

ber and chilly nights brought the larvae practically to a standstill. Prevented from feeding twenty-four hours a day, they developed much more slowly than had their summer forebears.
Besides, it was clear that their enemies had multiplied during the
warm months, so that even though I managed to ban stinkbugs
from the nets, those predators now patrolled outer surfaces of
such containers in profusion and I frequently found a bug busily
sucking a caterpillar that had come within range of a probing
beak. It amazed me that larvae could be successfully speared in
this manner. I would have thought that they would be able to
retreat, but apparently the bugs either were able somehow to
hold to their victims, or else they injected a paralyzing substance.
Well, I couldn't fight a whole army of bugs on the outside, so I
helplessly watched caterpillar numbers dwindle and simply
hoped there would be enough survivors to give me an answer—
even a computerized answer would do at this point!

The slackened rate of development, plus greatly increased
predation, made me wonder very seriously what the advantages
to luna of double-broodedness might be. In the cocoon box I had
a crop of Prometheas that had been reared only a week later than
my luna siblings, but the Promethea cocoons were slumbering in
deep torpor and would continue in that state until next spring. If
lunas had been content to do the same thing, would they not be
better off? I had garnered 126 hatched moths from the first luna
generation—from a single female—but it now looked as if I would
be lucky to get half that many from this second attempt with
more eggs to begin with and with three females providing
them.

The answer—several answers—of course, was that my doings
were unnatural ones, having nothing in common with wild lunas.
The fact that a second brood appeared was proof enough of that
pudding: It must have value to the species; the moths would not
be endowed with that characteristic if it were a harmful or perhaps even a neutrally useless one. The trouble was that, for all I
have said about lunas, I really knew nothing basic about them. I
am in the midst of telling that predators began to claim my attention in a painful sort of way (since I did not worry about them
so acutely during the earlier brood's growing up), but what about

parasites? I had excluded those, it seemed, or else they weren't as enterprising or as abundant as predators, for none either sneaked into a net or attempted to lay eggs on the larvae through it, so far as I ever was able to detect. Yet wild luna caterpillars probably suffer a high mortality from the merciless internal borings of wasp and fly larvae after eggs are deposited on or in them by the winged harpies. Predators—birds, stinkbugs, mantises, wasps, and other predacious insects—are likely to locate caterpillars in a hit or miss manner, simply by searching and, with the exception of 'a few distasteful kinds, all caterpillars are welcomed, so that lunas are not special targets in this instance. But parasites are frequently much more narrowly host-specific, a particular kind of wasp or fly often requiring a very special kind of caterpillar for its future larvae. Sometimes that means that the caterpillars must belong within a single lepidopteran family; or more narrowly, to a certain genus; and, most specialized of all, to only one species within the genus. In cases of the most limited spectrum of suitability, the parasite will be the more highly adapted, with sensitive receptor organs for locating the specific kind of caterpillar it requires. If such parasites exist in luna's case (and that is something I don't yet know), one can picture them flying through my woods, after predators have reduced the caterpillar population by consuming those *easily found*, using their diabolical antennae to follow down the scarcely discernible whiffs emanating from a few remaining escapees. It is the few survivors of this second hunt that become parents of the next generation. A moth like my Wine Female Number 80 that lays 685 eggs as good as talks to you about the vicissitudes of being a wild luna caterpillar.

For me to know the truth about lunas is beyond anything possible for a human being to do. I should have to take those 685 eggs and set them out in small numbers of one to perhaps no more than five, high and low, on persimmon, sweet gum, walnut, and hickory leaves. Not one must be protected by a net. Then I should remember where each one is and follow it through its life history, observing the kind of fate that overtook each of them in turn. Besides the patency that this could not be done is the equal obviousness that if it *were* done the situation would become abnormal. Birds would be frightened by me from their hunting, for example, and caterpillars that should be eaten under natural conditions would survive.

Therefore I must settle for the little bit of artificial knowledge I can gather from captive specimens. And if I were asked why I wanted to know, I would have to confess that my curiosity is based first on luna's beauty. It would be equally profitable to study carrion beetles or crickets, and I would do that if they were the only insects available. Who knows that they may not be easier to handle and have even more instructive lessons to disclose? But luna *is* there, and I *am* free to choose, so I allow the cool mint-green tint to make up my mind.

Caterpillars dwindled in numbers so rapidly that on September 12 I was able to record, "Seven dwarves present!" with a sense of exultation. Offspring of Wine Female Number 19 were not in much better supply. Things looked much more salubrious in Wine Female Number 80's net, but the hatch in that one had, after all, been 602, which is an awful lot of caterpillars! How can you tell the difference between 602 and, say, a mere 302 small green objects? With human beings' *advanced* capacity to see only five or six entities at a glance, we can't approximate hundreds very readily. These caterpillars, moreover, had hatched over nine different days, though the majority emerged in the first four or five, and it was easy to miss some of the smaller ones. I knew, however, that there were plenty of them—no doubt about that! Despite my slaughter of stinkbugs both inside and outside the net, the leaves were crawling with larvae and soon became badly chewed, even though the largest babies were still of a tender age. I realized that there were many more than I needed for my experiment, as well as more than enough to keep me busy were I to attempt rearing them all. Consequently I chose somewhat over a hundred at random (this included various sizes) for transfer to a new branch within a protective net and gave the others up to predators and stinging parasites. So far as this last act was concerned, I felt like a real heel, because those helpless larvae were concentrated in one place, not disposed in wide dispersion as their mother would have distributed them in her ovipositional flight. They were "sitting ducks": A parasite might be confounded by such an aggregation and be unable to infest more than one or two, thus wasting what must loom as a never-to-be-repeated opportunity; but a predator would be at no loss whatsoever. I knew that I might still see the caterpillars on several ensuing days, but eventually a morning would arrive when something would have made a clean sweep of them all and the branch would be bare. Perhaps it was in the nature of these caterpillars to feel uncomfortable in their crowded condition, for as soon as I removed the net some of them began to vacate that branch and move to neighboring ones. One or two of those might escape, casting another evolutionary vote for the wisdom of living a solitary life if you are a tasty caterpillar.

During the early stages, or instars, I could not tell whether the dwarf larvae were or were not of less than average size. Nor were any of the Number 19 or Number 80 caterpillars wine-colored. But as for the latter, their mothers had been green in their early instars, too. It was only in the last stage, preceeding cocoon spinning, that they had acquired their unusual coloration. Therefore I had to wait for that culmination of larval life of all three batches of descendants. And that gave me a great deal of worry, because not only were the cool September nights now slowing down development, but I was facing an October 15 deadline when I was leaving the farm to make a study of bird populations in Turure Forest for the Trinidad Regional Virus Laboratory deep in the American tropics. (Retirement on a farm in South Jersey? Yes, that was still true. I was going to Trinidad only as a temporary consultant, which I consider compatible at least with the pretense of retirement. Anyhow, I don't want to argue about it.) I urgently required all cocoons to have been spun, weighed, and put to bed for the winter in numbered cubicles in my cocoon boxes (I had had Rudd build three additional ones) before that date, though there was no way to hurry the caterpillars if climate and their natures caused them to dawdle. I could instruct Rudd in taking care of some of those last tasks, probably omitting the weighing, but it would be terribly tantalizing to know only some of the answers before my departure. Therefore I prayed for warm weather, while instead the days and nights turned prematurely colder than ever, and hurricane Doria brushed past one night giving one and all an impolite jostling that may have done some harm but assuredly no good.

It is astonishing how many things you can worry about if you put your mind to it. Now I began to wonder about early frost. What effect would that have on caterpillars? Of course it could be bad for them indirectly, by killing the leaves that they should still be eating. But would frost be lethal to the larvae themselves? I remembered that the pupae were frostproof or winter-conditioned, or whatever the correct term may be, but that was no assurance that caterpillars were similarly protected against low temperatures. Leaves and buds were much the same things but had different cold weather tolerances. The cocoon itself was no shield against a freezing environment, for that thinly woven structure had no appreciable thermal insulating properties. The pupa itself, rather, was adapted, as a part of its dormant condition, to survive whatever rigors winter chose to visit upon it. (In that case summer and winter pupae might be different from each other, just as the moths they produced had either purple or yellow wing edgings. That would be another interesting study that one could easily make with a deep freeze in July and a ticket to Florida in October—indeed, perhaps I ought to take a few cocoons with me to Trinidad.) I suspected that caterpillars might not be similarly endowed, in which event a heavy frost might be as tragic for them as for foliage. Then, not sure whether this was wishful thinking or not, I remembered that South Jersey is really in a different life zone from Philadelphia, with affinities to the coastal Carolinian region. That must be why double-broodedness is possible here, and *ergo* the first killing frost must be delayed long enough to bring all participants to a happy ending—perhaps even up to November.

As September passed into October the leaf problem became a truly respectable worry for which I did not have to make excuses. In fact, respectability carried with it a dull and fully conservative importance with no attending frivolity or glamour to make it interesting. Dogwoods, sassafras, and sour gums began putting on tones of crimson, with here and there a dash of yellow, but persimmon simply began to look sick and worn out. Many branches had all their leaves heavily infected with a kind of small punctate gall which forced caterpillars to eat around the tumorous places. That was not very difficult for them to do, because their orientation to a leaf was just the opposite to ours when we eat a sandwich: Their jaws work from side to side and they advance upon leaves edgewise, as if we attacked a sandwich held against the forehead, nose, and chin. Throwing their heads back in extension, they chomp-chomp forward and downward in a semicircle, taking a row of a dozen or so dainty bites along the way. When a leaf was perfect, they gradually consumed it entirely in this fashion. Galls caused them to waste a lot of food, either because they abandoned a leaf while it still afforded some good feeding areas, or because the galls put them off course and made them prematurely cut off the remaining edible portions. For as a rule they severed tatters of a leaf when they were finished with it by chewing through its stem near the twig. Consequently the conspicuousness of their ravages was lessened and, as they went on to new leaves, they were probably harder to find by questing birds than if they had left their picnic grounds littered.

Persimmon leaves also became markedly blackened on many trees. I don't know what caused that discoloration, but it seemed not to be any sort of external deposit, whether produced by a superficial fungus or by a gummy extrusion from within. It was smooth and would not rub off. At any rate it was definitely not a fall color! Some few leaves did turn yellow, though those appeared to be simply old rather than going in for autumn sports. My caterpillars did not like black or yellow leaves either, so net changing became a more frequent chore simply because a fully laden branch nevertheless no longer bore as many eligible leaves as it had in summer.

I gave no thought whatsoever to switching the caterpillars to sweet gum, which was both abundant and as yet largely unresponsive to the autumn chill. What I have said about caterpillars' preferring to feed on their mothers' host plant is even more true of an individual caterpillar during its single life time: It would have an absolute horror of changing to anything else, once it had started on persimmon. Perhaps I could have induced my larvae to do so, but even if the effort had succeeded, I would have deplored the need for such an upheaval, for it would now be impossible to make full comparisons between this brood and the summer one. The change in diet might of itself have no influence on what happened (though I could not be certain of that either), but some loss of time must be incurred while reluctant worms became at last so famished that they ate hateful sweet gum leaves in desperation. And what would be going on while they thus languished on their hunger strike? Why, they would be aging just the same, so that when spinning time arrived they would not be as large as if their feeding had been continuous. Therefore I could not validly contrast their weights with those of the first sibling generation.

But like the children of overanxious parents, those caterpillars *did* grow up and no amount of my worrying was enough to retard them more than nature intended in this late season. As they molted into the final instar I began looking eagerly for the first signs of a shading with rosé wine in the Number 19 and Number 80 nets. The more closely I scrutinized the larvae, the more uneasy I became. It was not that I failed to detect wine-colored individuals—I was prepared for that change to take place during the last few feeding days, for that was when I had noticed it before. But these caterpillars struck me as being somehow different from the summer ones, and I was uncertain whether my impression was the result of nothing more than closer observation now or whether they really did wear a different guise, as the mature spring and summer moths had differed from each other in their purple or yellow outer wing borders.

The trouble was that I had not taken elaborate notes on the appearance of last generation's caterpillars. To call them green with a narrow yellow stripe along each side had seemed an adequate description to me. But among the present brood I began to notice all degrees of a variation that *ought* to have impressed me enough to remember it even without notes, and I could not honestly say that the trait had registered on my consciousness! This had to do with markings of the tubercles. Caterpillars, especially those of the silkworm group, are really sullied when called worms, because their segmented bodies are hardly so simple as mere smooth cylinders. Their segments are more like octagonal blocks, and at the apex of most corners of each octogon lies a small hardened prominence called a tubercle. There are thus four rows of tubercles running along each side of a saturniid caterpillar's thoracic and abdominal segments (almost a dozen in all), and depending on the species of larva, its tubercles may be knobbed, spined, tufted, pigmented, or ornamented in some other fashion.

One of the characteristics of lunas seems to be the ability to form brilliant pigments in practically all their stages. That property, which I originally thought pertained only to the adult state, now began to manifest itself in caterpillars' tubercles. These markings varied from tiny, almost negligible dots in some individuals, to conspicuous eye-catching discs in others, and their shades likewise ran from light pink through brilliant coral to a crimson so dark as to appear almost black. Thus some caterpillars would still scarcely draw my attention from the standpoint of those decorations, whereas the outstanding ones almost matched green persimmon leaves studded with escharotic galls.

That lapse in my memory bothered me until I forgot about it with the advent of a far more serious one—a lacuna so great as practically to invalidate my entire study of third generation luna moths. My scrutiny of tubercles (and I should say that "black face" and "green face" constituted other traits that I began belatedly to document)—my attention to tubercles did not sidetrack me from the main quest which was for the earliest signs of wine-color suffusing an entire body. When I had almost concluded that the larvae were destined to remain a wholly green population, the two females—Number 19 and Number 80—having failed to transmit their vinous heritage despite sibling inbreeding, a deeply red larva appeared in one of the Number 80 nets as if by an almost instantaneous change. All I knew was that I had changed that net to a new branch a day or so previously, examining each caterpillar carefully as I transferred it, and there had not been the faintest indication that any of them was tending away from greenness.

Well, I could not argue with this one: It was as deeply colored as I could have wished, and moreover it was crawling about the net as if ready to spin. Therefore I removed it to a jar in the dining room, giving it a sprig of leaves in case it should be hungry, but it used the sprig immediately as the supporting framework for a cocoon. That did not astonish me, for once caterpillars begin to crawl about actively they have usually finished feeding, emptied their intestinal tracts of the last remnants of waste, and are now possessed of a blind wanderlust that leads them into all kinds of places, good or bad for cocoon spinning as the case may be. This is another of the points in their lives when dispersal is achieved, though I imagine it serves mainly to spread cocoons apart. If all caterpillars climbed down tree trunks and spun immediately on reaching the ground, scavenger-predators such as skunks would have an easy time finding them simply by rooting around one bole after another.

Next day another red caterpillar, also in the "crawling" stage, appeared in a Number 80 net, and on the following morning I found not only two additional ones in that situation but also a fine "wine" crawler in a Number 19 net. In short, it now seemed that wine was nothing but the tint assumed when a caterpillar was ready to spin. Some metabolic change occurred, probably in its blood (for the color seemed to come through the translucent skin), when feeding had terminated and the gut was emptied. A larva, now reaching the ground, was actually protectively colored among dried oak leaves and some recently fallen deciduous shards still bearing remnants of brilliance. Nature thus provided for safety of the late brood by turning them all to wine, while summer ones sought the ground as green crawlers that would blend with the prevailing verdant cover of that season. A wine-colored larva in July was therefore only a slight accident, a miscalculation in a metabolic machine here and there among a multitude of brethern who performed without defectiveness. Similarly, green would now be relatively disadvantageous, and as individuals in the two batches successively abandoned feeding and became crawlers, I found remarkably few that assumed only a faint tinge of red: The rule now was to drink deeply of October's vintage.

It flashed upon me that in my teens at Andover I had seen much the same thing. Arriving in Massachusetts for the fall term in September, I would find the nights already quite cool and the campus littered with elm leaves that formed insubstantial wind-rows with each changing gust. Among those castaways were a few similar-sized objects which did not blow about but clung miserably to walkways, many of them having been stepped on. These were a kind of hawkmoth or sphinx caterpillar that, in a natural woodland, would have burrowed underground to pupate for the winter. Here they could obviously not make their way through asphalt, while even the closely mown lawns seemed impenetrable to them. The point of my recollection is that most of those larvae were wine-colored, a datum which then impressed me only of itself, not with a notion of its function. I don't remember whether I thought that the caterpillars might have been green in the elm trees and had turned red only on crawling (or falling) to the ground, but I do so conjecture now. In any case, it is a helpful happen-so in my experience, for one can generalize more easily from two samples than from only one.

As further confirmation of my sudden conviction—or enlightenment—about the seasonal aspects of wine, I now found in my summer notes a passage that had not seemed significant then. On one of the days that I removed a wine-colored crawler from a net (and, too, it now occurred to me that I had never seen a wine-colored caterpillar *feeding*), I was forced to disturb several larvae that had already begun to spin in the neck of the net where it was gathered together and tied about the branch. My comment was simply that those caterpillars had not yet gotten very far along with their work, and would now have to begin all over again, but it would not inconvenience them very much. Period! Does this not indicate that they were *green*? Since I had disrupted net affairs in quest of a wine-colored crawler, I must surely have noticed if spinners, partly concealed, were suddenly exposed as wine-colored themselves.

Incidentally, I found another item in my notes that now bothers me: dark green caterpillars. Two of these were seen, several days before the first crawlers of any color appeared. Unfortunately I did not follow them meticulously. I drew the inference, almost certainly a false one, that these had preceded and become my two winos, but of course I could not vouch for it. Perhaps that is a trait I should have been following, rather than the false rosy lead.

But here I was arguing myself out of the very result I had tried to bring about! The trouble was that the result was *too perfect*. Biological processes are rarely uniform, as one has only to remind one's self by reflecting that the kernels in a pan of heating popcorn do not all pop simultaneously. Wouldn't we be astonished if that should happen one day? We expect, rather, the early-bird popper, then a few somewhat less early ones, followed by the general rabble of "Me, too" poppers, and finally a decreasing incidence of late kernels down to one final individualistic "pop." Therefore two netfuls of largely wine-colored larvae immediately made me look for some other reason, other than my hypothetically imposed genetic one, to explain the phenomenon.

Give me a good quantitative character to work with, not color! I suppose that if I had taken kodachrome pictures of each sibling larva at maturity last summer, I could now tabulate all their traits, black or green face, tubercles, and even depth of green body by use of a standard chart. That might enable me to salvage something out of my "wine" experiment, but from standpoints that I had not considered originally. Now, however, I still had my dwarf strain to play with. Here was a gorgeously concrete problem. You weigh your cocoon; you pick up the moth and measure it with your ruler. You enter the figures in a book, and however you decide to manipulate the numbers, you don't have to call on vague impressions, trying to remember whether a specimen weighed this or that, or spanned one distance or another.

Come to think of it, the dwarves would give me one final check on the wine question. If they all went into their cocoon hibernacula in brilliant green, my arguments about autumn protective coloration would crash and the genetic buildup must be resurrected. But I did not have even hopes for such an outcome. It was good enough for me to have learned, however blunderingly and with multiple false premises, that colors of summer and fall caterpillars were adapted to their respective seasons. Let dwarves turn as red as they chose: I was more interested in the sizes they would attain.

Bugs had eaten many of the dwarf babies, and I continued to find more bugs in the net from time to time, so that I ended up with only fourteen caterpillars. To be sure, this was several more than I thought were present at the lowest ebb of my optimism, but it was still so few as to be marginal where the statistical validity of potential results was concerned. They would have to put on a striking show to be convincing. In other words, such a small number could not very well string itself out like the popcorn kernels, to demonstrate both the outer limits of size and the clustering of average individuals in the middle. Suppose that one dwarf caterpillar grew to very large size, while the rest remained below normal: That one, in a series of only fourteen, could indicate that if the bugs had spared another fourteen larvae, additional intermediate individuals would have appeared, and by extending this thought one could come to the conclusion that the dwarf progeny displayed no more than normal diversity of size within the luna species. Thus: this experiment negative also, or *Q. non E. D.*

Only a few more days must elapse before I would learn all the answers, for the dwarf larvae should begin to spin very soon. Trinidad was really breathing hot down my neck now. Some details were certain to be left dangling, but enough specimens were popping to supply the substance of many solutions before October 16. Earliest eggs of my three breeding females, numbers 19, 80, and 99, were only three days apart in age, so their first spinners would not be likely to show a much wider spread in time. They *were* slightly out of step already, since first spinners of Number 80 and Number 19 had put on reverse performances from their hatching order. I realized now that Number 80's caterpillars had had the sunniest exposure and may have been speeded up (or correspondingly not retarded) thereby. The seasonal influence is most apparent when we look at the interval between hatching and spinning of the matriarch's progeny in July: The shortest interval was twenty-six days, while in the case of Wine Number 19 progeny, developing mostly through September, it was thirty-eight days or almost half-again as long. The first

cocoon, on October 2, was coincident with my driving to Sears
catalog sales store in Wildwood to pick up my new field clothes
(for I refused to appear in my old rags on foreign soil). Hurry,
hurry, hurry!

Already I had a most interesting observation on cocoon
weights—*un*corrected ones, that is. I could now confidently believe that the earliest spinners were males. Therefore a comparison of, say, the first eight (I lumped progeny of Number 19 and
Number 80 together for this purpose, since they could both be
considered of normal size) with their hastiest eight uncles should
be a reasonable sort of lineup to attempt. Here I at once could
see that the fall crop had not only taken longer to mature, but
had also attained a smaller size than their summer predecessors.
The average fall male weighed only 3.6 grams, with a spread in
this small series between 3.2 and 4.0 grams. The summer cocoons
had ranged from 3.9 to 5.1 grams, the average being 4.4 grams.
No computer was needed to demonstrate that these figures were
"significant" in the statistician's significant sense of that significant
word.

But that was far from the end of it! I selected those eight
early cocoons because that was all I had up to that time. I was
so eager to analyze my new results that I was doing it on a day-
to-day basis. Twenty-four hours later, when Number 80 larvae
were popping into wine-colored spinners (with, yes, one or two
almost unadulterated green ones along the line), I harvested two
dozen additional cocoons, among which only three attained 4
grams or fractionally more than that weight, while two tipped
the scales at a measly 2.6 grams. *This was 0.3 grams lighter than
the fresh (uncorrected) weight of my two summer dwarves.*
Thus it appeared that the progeny of Number 80, a moth who
by this time could be regarded as a grand old lady with her gargantuan ancient weight of 6.3 grams, were not coming up to
maternal standards at all, but, rather, were distributing themselves around an entirely new norm as regards mass. In such a
case my summer dwarves were *not* genetic freaks but, like the
two summer wine larvae, no more than out-of-season expressions
of normal luna variability. The only aberrations I had missed so
far, then, were purple-edged second-brood moths, or yellow-

edged spring ones. The way things were going, I had no doubt I should run into those anomalies, too, if I kept on raising caterpillars long enough.

Seasonal colors, sizes, weights—what did these mean? I have already offered a hypothesis to account for imagined advantages to larvae in being green in summer and rosy in the fall, but I can't think how purple edgings may benefit spring moths compared with yellow borders in the hot-weather brood. Some weeks previously I had more or less bemoaned the summer brood from the standpoint of heavy mortality suffered by larvae at the jaws and stings of predators and parasites. But now I began to piece together an interpretation. Perhaps my arguments are too facile. After all, I am backed up by the facts of what happened, so I do not have laboriously to build a basis for talking about it. Hence anyone could say what he liked and remain unassailable so long as he stayed within the boundaries of the uncontested data.

So, here goes. If we hark back to my Number 80 female and remember her as the author of 685 eggs, of which 602 were known to have hatched, we can salute her once more as a great bulwark against the extinction of luna moths in my woods, for 602 dispersed caterpillars are a mighty large challenge to hunters that would seek to exterminate them all. Almost any moth life insurance company would write an inexpensive policy as regards the likelihood of a few of those caterpillars' surviving—it need not be many, but only just enough to provide breeding stock for the next generation at a level that would keep the population constant. For, mind you, we are concerned to prevent extinction, but it is also similarly important to avoid overproduction. A plague of caterpillars would strip and kill the trees, would afford parasites an opportunity vastly to increase *their* populations, and in the next year the erstwhile teeming lunas might really be starved or stung out of existence. Therefore I am considering only the maintenance of a stable number of moths, allowing some variation, of course, since nature is rarely at a motionless equilibrium, but in essense a succession of years which in luna archives could be listed under "Business as Usual."

But Number 80 *did* emerge when she did, instead of sleeping through until next spring, and her larvae, if unconfined, would have had to face the dangers I have enumerated in their late summer enhancement. Or *would* they? This is the question, in whose answering I derived a notion of the function of that summer brood. The moth, upon emergence, was confronted immediately not only with the biological needs of her position—mating and the scattering of eggs—but also with the provisions of that life insurance policy. The cagey agent had not written in a double indemnity clause in case of accidental death. Larvae face all kinds of risks, but so do adult moths. We have already noticed what a Cardinal will do to a fat female luna. Suppose Number 80 had been snuffed out (or snipped apart) before she had mated, or perhaps after mating but before laying her first egg? A complete loss, total tragedy!

On the other hand, her survival until she had laid some of her eggs—let's give her the benefit of half of them, which would keep Cardinals in the background for only her first two days anyhow—longevity for even that short time would amount to a kind of insurance of the not-all-in-one-basket type. Even if she could be represented in the next generation by only small daughters, they would help luna matters by requiring separate Cardinals, or one extraperspicacious Cardinal, to do them all in. Three such descendants, each containing only two hundred eggs and each living long enough to lay only half of them, would accomplish as much as the original Number 80 had done, *granted she had survived even her first two days.* But let her be wiped out before that and all would be lost. Let one daughter, or two if need be, meet premature destruction: A descendant will still remain to carry on the family strain.

The fall brood is thus a means of fragmenting energies that had become concentrated in large moths that were able to attain great (and conspicuous) size during the most favorable growing weather of the year. Apparently this is useful, for otherwise one would not observe it. However, it may not be *vital*, for other moths such as Cecropias seem to get along eternally in Eldora, New Jersey, on the basis of one yearly generation. But the frill of double-broodedness, if one is tempted to regard that trait as some kind of luxury indulged by lunas and some other forms, may be more than that, the question in my mind being only one of the degree to which luna *must* depend on such a feature in its life history. Like the colors of wing fringes, this function remains interesting and of obscure value.

October 9! A week from this morning I would drive to Avalon, park the Buick in the cellar of our shore cottage and walk to the bus station en route to New York and Port-of-Spain. Oddly enough, that seemed the best way to begin the journey from Eldora. On that morning I would leave the farm at daybreak, so that all observations on caterpillars and cocoons must be completed by the preceding Sunday evening or else be forfeited. Then imagine how avidly I look into my nets on this depressing rainy ninth of October, realizing that the minutes are now slipping away as rapidly as ever, while clammy air conspires to slow caterpillar reactions to ever more sluggish rates!

I had changed the dwarf net to a new branch yesterday. Not that leaves were running short—there were too few caterpillars in this net for that to become likely. Indeed I had not changed the net since placing hatching eggs in it. It was the very fact that leaves had been consumed so scantily that led me finally to decide on the shift. I could not make a satisfactory count while larvae had a chance to hide among all that foliage, and I simply would not defer the urge to make an absolute count of dwarves. As already stated, I found fourteen. Eleven of them looked mature to me. All had been incontestably green.

Now, still on October 9, I suddenly burst into mad cackles of glee. It was most fortunate that no one was on hand to hear, for it would have been difficult for me to explain why two *crawling, deeply wine-colored* luna caterpillars were sufficient cause to evoke that kind of reaction in me.

And then, in the remaining six days, everything else finally fell *out* of place, too! In the first instance, one of the dwarf cocoons, at 5.2 grams, was not matched by any of the seventy-six cocoons of the great Wine Number 80 female's children, or by the thirty-seven progeny of the more "normal" Wine Female Number 19. All those were lighter than that, indicating for one thing that no whopping big females had been produced in this generation.

But one would still like to know why dwarves should outstrip the average (for the other dwarf siblings also were above the present normal) unless out of fateful perversity. I believe it had to do with noninterference. Changing the nets must have bothered caterpillars much more seriously than I would have supposed likely. Nevertheless, let's argue that that was not the reason, either: If I can't give the explanation, is there still something that I can conclude about the dwarf trait? As I have already suggested—as I feared might be true—there never was any such trait, and consequently I could not enhance it by inbreeding.

As I reviewed all my results, in the light of the latest and final information entered in my notes on October 15, I admitted to myself that I had accomplished exactly nothing. Heaven help the Trinidad Regional Virus Laboratory as I now bent my mind to *its* problems!

Chapter 4

THE CRESTED FANTAIL

Pigeon love is an emotional condition quite distinct from the generalized positive response to birds that draws certain people to the study of ornithology, either as amateurs or professionals. Indeed some pigeon fanciers act as if they are hardly aware that other kinds of birds—and bird students—exist. They are all extremists, so far as my observations have led me to conclude, and but for a careless or, perhaps, dishonest clerk in a pet store, I would have become one of them very early. That man sold my father an alleged mated pair of white Fantails, which I received rapturously for Christmas when I was nine. Father did not know, nor did I, that the birds were both females. I kept them in an abandoned chicken house where my twenty-three box turtles had reigned the previous season, hoping soon to people the cote with baby Fantails. But when spring and summer passed squabless, my briefly aroused pigeon ecstasy waned and box turtles resumed their whitewashed throne.

My father was a persevering man. On Christmas when I was ten he gave me a pair of blue-barred English Pouters. I did not find these awkward, feathered-footed, excessively slender birds nearly as appealing as the chunky, chesty, head-bobbing, pirouetting Fantails. However, they *were* a proper pair, and I suppose I would have become converted to them had not the female escaped and been run over by a delivery truck. Meanwhile one of the Fantails had died. Now the male Pouter married the remaining female Fantail and they had a hybrid squab. I remem-

ber that bird vividly, since it had some of the traits of each parent, though all of the characteristics were intermediate or dilute
in form. Its only moderate slenderness, scantily feathered toes,
spade-shaped tail, and indecisive neck shaking did not recom-
mend themselves to my enthusiastic devotion—I could see the
youngster merely as a curiosity. While I continued to keep pigeons
for several years, the fanatic streak they ought to have aroused
in me remained dormant, whereas I was rapidly developing into
a conventional watcher of the native wild avifauna.

In adult life I reverted to pigeonkeeping from time to time,
but always with ordinary stock, that is, the mixed blue and
brown birds with variable splotches of white in the wings or on
the body that one finds in a semiwild state in cities and on farms.
But all the time I sensed that these were only substitutes for the
birds of my dreams that might some day fulfill an aborted childhood vision. When at last I acquired my South Jersey farm, I
knew that it was time to dredge up the disappointment of half a
century ago and confront it with a real pair of Fantails. If I could
now face purebred squabs with equanimity, all should remain
tranquil; if not, there was a chance that long-suppressed mania
would supervene.

Although Rudd Kolarich knew all kinds of useful things, I
would not have blamed him for flunking my Fantail question.
However, he passed it easily.

"Go and see Fred Kludzuweit," he said. "A couple of miles
south of Cape May Court House you'll find the Country House
Restaurant on the Shore Road. That's Fred's new place, and he's
got all kinds of fancy pigeons, chickens, ducks, turkeys, and
geese, and even peacocks, in cages back of the parking area. People take their kids there to see the birds and then eat, or maybe
it's the other way 'round. Anyhow, the birds are a great attraction
and if Fred hasn't got Fantails, he would surely know where you
can get some."

I went and looked. Rudd had not overstated any part of it.
I saw breeds of pigeons that I did not dream lived on this planet.
There was one pair, indeed, whose race must have had a most
tenuous hold on survival. These had great rosettes of feathers on
each side of their heads, so that their eyes and even their beaks

were entirely concealed. It so happened that they were attempting to mate as I stood before their cage, but the male—having mounted the female and being unable to see where he was—had become disoriented and was trying to copulate with his wife's face. How can organisms as handicapped as that propagate themselves? I suppose that half the time—strictly by the laws of chance—they fall into a correct fore and aft alignment.

Rudd had told me to march straight into the kitchen through the back door and make my wants known. I would like to have done so but simply did not have enough brass. Anyhow, wouldn't that be the best possible way to make myself unpopular? If I had been the owner, I would have thrown such an intruder out before hearing what he had to say. Besides, I had not seen any white Fantails in the cages. There were some browns and blues, as well as a pair of blacks, but to exorcise my latent demon I must have white ones. So why bother the man?

On the next day Rudd took me in hand and we "marched straight into the kitchen through the back door and made my wants known!" Fred Kludzuweit couldn't have been more cordial. He was sitting behind a huge mound of cole slaw, but kept on chopping up more cabbages as we talked. He said he had a farm nearby where he kept most of his birds (as well as horses). Just at present it happened that he had no white Fantails, but he would ask one of his friends in Wildwood Crest to send over a pair. If I'd come back in a week, the pigeons would be waiting for me.

That was just enough time for Rudd to fix up an indoor-outdoor enclosure—an inside nesting and feeding area, communicating with an outside flying cage—under my direction. There were several sheds on my farm, one that had formerly housed sheep and, later, chickens, during Rudd's tenure, being readily adapted for use by my brand of livestock. The requirements I now specified related chiefly to ratproofing, for I had eventually lost my remaining childhood Fantail, as well as several common pigeons, to bloodthirsty domestic rodents. But I felt that the flying pen should be hawkproof also in these Cape May County wilds where the accipitrine or true bird hawks—Cooper's and the Sharp-shinned—were fortunately still common enough to worry about, especially during the fall migration. Eventually I realized that all of those precautions were effective also against opossums and skunks, if not raccoons besides. Indeed, what about weasels, foxes, stray dogs and cats, Great Horned Owls—where did the list end? Perhaps a farm was not a safe place for pacific pigeons after all, if enemies were lined up as thickly as that. I checked and rechecked Rudd's heavy metal screen wire and sheet tin work, until I was sure there were no holes wider than a cricket.

I could tell at a glance—even when they were still in an alien cage at Fred's parking lot—that these Fantails were mates, for one was nibbling the tiny feathers about the other's eyes and beak. You don't do that to people you don't like, or, from a reciprocal viewpoint, you don't *let* people you don't like do that to you. While it occasionally happens that two pigeons of the same sex will form an attachment for each other that has all the outward signs of a heterosexual mating, such behavior is not common, and anyhow I felt the odds were strongly against my having the same bad luck twice. As a matter of fact, my two original female Fantails had not put on a lesbian show (which might have excused the clerk in the pet store)—apparently only exceptional birds find it possible to engage in that sort of *pas de deux*.

But now was no time for rumination about what had happened or what might still happen, because I became occupied immediately with what rapidly did begin to go on. My Fantails were "birds of last year," according to Fred, and that should put them at present in their most fertile and prolific age. The date was June 2, surely a classical time for reproduction. Meanwhile they must not only become acquainted with their new quarters but also forget about hankering for old ones, as pigeons do, before they could settle down to unharried lovemaking. The urge of youthful maturity drove them to make those adjustments in the minimum time. Exactly twenty-three days later egg Number 1 appeared in the nest. On July 13, 1967, my father's design was realized at last when I lifted up the gently protesting female and beheld a pair of newly hatched squabs beneath her hot breast.

At that instant I very clearly and definitely felt something suddenly go "thunk!" inside my head. The devil was still active in there, and now that I had given him something to work on, he was about to take charge. Oh, well, I had already had a good life. When something went "thunk!" again almost immediately, I realized that he was talking about those Pouters. That experience, too, must be recapitulated. Very well—after enjoying my pristine Fantails for a while, I would branch out into some crossbreeding experiments.

The first set of babies were scarely half-grown before a second clutch of eggs appeared. If it was white Fantails I wanted, I was now going to have them galore. I could hardly contain my elation as weeks passed by and I would see four, then six, and then more, lovely white Fantails sunning themselves on perches in the flying pen. The interior compartment became crowded. As the first babies became sexually mature, fighting broke out between the father and his sons, and later between brothers. Something must soon be done to control the flock's exuberance.

But it was not done soon. Instead, I went off for four months as consultant to the Trinidad Regional Virus Laboratory and then treated myself to an additional month of birding in the hinterlands of Peru. When I returned to the farm on March 27, 1968, I found that I possessed twelve Fantails, of which four pairs concurrently had eggs. The reason there weren't more, according to Rudd, was that only the original birds were fertile. Thus my flock consisted of two parents and ten siblings but no grandchildren. The explanation for that situation was a simple one. In any small social group of pigeons there will be a dominant male that persistently interferes with copulation by other males, even when those subordinate ones are trying to mate legitimately with their own wives. Since Fantails find it physically difficult to mate under the most serene circumstances because of their spreading and unwieldy tails, my master male would easily have prevented his sons' successful fertilizations. Thus he had saved me from being richer than I already was.

In a scientific experiment you sacrifice your outworn or surplus animals. To be scientific, then, I must simply wring some of my Fantails necks. However, my secret devil had so corrupted the objective researcher in me that I now loved even the weakest Fantail of them all too much to put my hand on it as executioner or to deliver it to a deputy slaughterer. But how could I give so many pets new homes in which they would continue to receive love?

During my absence Rudd had asked an Italian friend of his to administer their first winter pruning to my young fruit trees. That gentleman had cut away expertly as a pure favor, but Rudd happened to mention that he had also admired the Fantails. Very promptly, therefore, I sent the Italian two of my frustrated young breeding pairs, which should actually have been a kindness to everyone concerned, including the overlorded pigeons. But that still left me with eight, and within a couple of months I was as badly off as before.

Now I must be drastic. I told Rudd that I would keep just two pairs of Fantails, while I would willingly trade all the rest for a single pair of some other breed. Remembering those bizarre pigeons with the rosettes of feathers on the sides of their heads, I suggested that perhaps Fred would consider making such a trade. By this time my surplus came to eleven birds (not counting the extra retained pair), which in exchange for only two ought surely to seem an advantageous deal to any pigeon fancier.

"I think they're called Nuns," I instructed Rudd as he set off with the crateful of bemused Fantails. "If he won't give you Nuns, ask him for something else—anything at all—though I'd much prefer Nuns."

It seemed to me that if I was going to crossbreed two strains, it would be the most interesting to begin with stocks as different from each other as possible. What I hoped to do—though I can't say when the devil whispered this to me—was to go beyond the mere hybrid stage and somehow work back until I had grafted the characteristics of one breed onto the other. In short, I foresaw a bird with facial rosettes *and* a fantail, though I did not really know how to go about achieving that result.

Rudd duly returned with Fred Kludzuweit's very best pair of Nuns, and if you have been ahead of me in this part of my tale, you will understand my consternation when I saw them.

THOMAS KINNEMAND, JR.

A "mosquito walk" on the farm. The author's companions are salt
marsh, or Jersey, mosquitoes—*Aedes sollicitans.*

THOMAS KINNEMAND,

The author's salt meadows. These are undiked, allowing small predatory fish to enter during tidal flooding. Mosquito breeding consequently is negligible here.

ALLAN D. CRUICKSHANK FROM NATIONAL AUDUBON SOCIETY

An adult Osprey or Fish Hawk tending its half-grown young. Such nests are used for many years. These birds have recently declined greatly in numbers, owing to destruction and pollution of their aquatic fishing grounds.

DR. C. BROOKE WORTH

Polyphemus caterpillars being reared in nets on a willow oak. The nets exclude most predators and parasites.

An almost fully grown Cecropia caterpillar. The spiny tubercles are harmless, though they probably discourage some would-be predators.

HUGH SPENCER FROM NATIONAL AUDUBON SO

THOMAS KINNEMAND, JR.

Luna moth. This specimen can easily be recognized as a male because it has wide feathery antennae and a small abdomen.

THOMAS KINNEMAND, JR.

A newly emerged male Polyphemus moth. A separate cubicle for each cocoon enables the author to keep track of the breeding performance and genealogy of individual moth families from year to year.

A newly hatched eastern box turtle, still bearing an egg tooth at the tip of its nose for tearing open the confining membranous eggshell.

DR. C. BROOKE WORTH

A Seaside Sparrow at its partially concealed nest.

ALLAN D. CRUICKSHANK FROM NATIONAL AUDUBON SOCIETY

A white-footed mouse on a rotting log in the woods.

JACK DERMID FROM NATIONAL AUDUBON SOCIE

THOMAS KINNEMAND, JR.

A Japanese mist net. The meshes become almost invisible when the net is spread open.

THOMAS KINNEMAND, J

Bluebird boxes at the farm are pre-empted by chickadees, titmice, Tree Swallows, and flying squirrels. Here is a mass of material assembled by squirrels, in which dozens of squirrel fleas, *Orchopeas howardii,* were hatched.

Here was another totally unfamiliar breed of pigeon, so far as I was concerned! Moreover, they were practically *normal* birds, their only unusual adornment being a semicircular band of feathers that grew in the wrong direction around the backs of their heads, pointing upward to form a low but attractive crest. This must be what gave them their name, though I now think that "Nurse" might be more appropriate than "Nun." Anyhow, Nuns they were, with pleasingly patterned plumage in chocolate and white—chocolate on tail, outer half of wings, chin, throat, face, and crown; white on body and nape. The design was neatly arranged so that the chocolate crown precisely met the white crest, creating a halo effect.

From the standpoint of crossbreeding, that would seem to present quite a number of separate characteristics to play around with, but I had not thought of color in imagining my experiments. I was interested only in form, and the Nuns afforded me only their insignificant crests in that department. If I may say so, I was crestfallen. For the time being I simply put the new birds in the coop with the Fantails and went about the more important chore of tending my caterpillar nets.

Of course the Nuns had no inkling of having disappointed me or of having to ingratiate themselves—they stolidly went about their pigeon business until, on November 11, they forcibly commended themselves to my approval by hatching out a set of eggs. They had established a small territory on the floor of the enclosure, an area that had previously belonged in its entirety to the subordinate pair of Fantails, and a small war had been required to effect the subdivision. Therefore revenge was in order. On November 22 I found the young Nun squabs pecked to death, presumably by the aggrieved male Fantail.

Belatedly I gave Fred the offending pair of Fantails—this making seventeen offspring that I had reared and sent out into the world. Now I expected the Nuns to repeat their nursery effort. However, they went into a heavy molt and then embarked on a long winter's rest, doing nothing but eating and looking pretty. The Fantails must have been ready for a breather, too, for they skipped at least one brood while molting along with the Nuns. The year 1969 came in with morale at the nadir.

Now I became piqued, for the last obstacle I had foreseen was uncooperative pigeons. Well, I would give them some stimulation (as, of course, would springtime). Whether they liked it or not, I would now wreck their marriages and inaugurate the new look in South Jersey aviculture. This would be the beginning of a quest for the Crested Fantail, and the sooner the first cross was effected, the easier it would be for me to plan subsequent methodical incest.

An immediate problem that I knew faced me was the constancy of mated pigeons' devotion to each other. Mates remain mated as long as they are together. Oh, a male will make up to any female in sight, and rarely he induces a strange female to let him copulate with her, but this by no means destroys the pair bond. On the other hand, if one of the pair dies or is removed, the bereaved partner will eventually accept a new mate. For a while I considered converting another small shed on my farm into a second pigeon cote, so that I could house the two Fantail x Nun combinations out of sight and sound of each other. That surely would result in the briefest period of mourning before wedding bells rang again. But such a maneuver would have made the whole project less convenient. Instead I had Rudd build a second indoor compartment adjacent to the first one but with a composition-board partition between them. The birds could still hear each other's characteristic cooing—even I could tell them apart by those sounds—but perhaps "out of sight" alone would be enough to do the trick. On March 2, 1969, I sent Fred my last three young Fantails (the total reared was finally twenty!) and set up twin experiments—male Fantail x female Nun, next door to male Nun x female Fantail.

The following day the male Nun had found a crack in the partition and was standing glued to the vision of his mate on the other side. A piece of adhesive tape fixed that, but I felt despondent about the chances for speed. "A little month!" says Hamlet. Now I must say, "A little less than forty-eight hours!" That is how long it took the male Fantail to console himself and the grieving Nun. I admit that the other pair waited for two weeks, which looked a bit more respectable, but that, too, would surely have shocked Hamlet. Soon both pairs had eggs and the experiments were in motion with all remembrance of the past washed out.

The four pigeons I had acquired from outside sources wore numbered, circular leg bands. I had not bothered to ring the baby Fantails, but now it was obviously essential that I be able to establish permanent identification tags for all the experimental progeny. Of course I had been banding wild birds for the Fish and Wildlife Service for many years, but it would not be legitimate for me to use any of their rings for private domestic purposes. The Fish and Wildlife bands were of a different sort anyhow—slit so that they could be pried open and then squeezed shut again when applied to a bird's tarsus. Standard pigeon bands, being complete rings, must be slipped over a squab's foot while it is only about a week old; after that it is too late unless you attack the baby within a day or two afterward and use liberal amounts of olive oil or saliva, preferably with a dash of profanity.

Not knowing where to get pigeon bands, I turned once again to Fred with my problem. Each time I had given him more Fantails he had sent back an invitation for me to visit his farm, but I had always been too busy—or thought I was—to follow up the opportunity. Now as I talked to him on the phone, he insisted that I must come so that he could give me some back numbers of the American Pigeon Journal which regularly advertised bands for sale.

Thus it came to pass that I stood with Fred Kludzuweit in the middle of a great roomful of pigeons of a myriad of sorts. Among them I saw some of those bizarre creatures with facial rosettes that I had thought were Nuns. I would not tell Fred about my booboo, but I did (with studied carelessness) ask him the name of those strange birds. Jacobins! Well, perhaps it had all come out in the most favorable way after all. These freakish pigeons might have been too delicate or simply too difficult to work with. If I could manage to breed crested fantails from my Nuns, that might already be a notable triumph.

The American Pigeon Journal would have been an amazing periodical to me if I had not been aware beforehand that pigeon people are the craziest. I quickly found a source of bands, but then I was arrested by a far more important advertisement. For many years—ever since my army days during World War II—I had been aware of the existence of a legendary pigeon expert named Levi whose whole life had been devoted to pigeons. The way he locked into the army was through his work with the Signal Corps in training homing pigeons to carry messages from the front lines when communications had been knocked out. In fact Mr. Levi had done this also during World War I, so that he seemed to be immortal. That he had written a book I knew, too, but it had not occurred to me that such a work would be of use to my superficial dabblings in pigeon lore. Now, however, I was getting involved in something deeper, and here providentially was notice that the book was still in print and readily available. On an impulse I sent off my check for twenty dollars—it had *better* be a good investment!

Talk about revelation: Chapman's *Homer* isn't nearly up to the same pitch. The 99″ x 11¼″ tome, entitled simply *The Pigeon,* and weighing four pounds, was all lean meat without an ounce of fat unless you are unkind enough to classify a color portrait of the author facing page *ix* as excessive. Of course I was delighted to be able to see what breed of person he might be, while it was an entirely unexpected fillip to find that Wendell Levi had inscribed the flyleaf with his best wishes and autograph. It was as if I had not only bought a book but also joined a club.

This was a true monograph and more—a work of love as well as of lore. Chapter titles give some indication of its scope: The Relationship of Pigeon and Man; Breeds and Varieties; Anatomy; Physiology; Genetics—Variations and Inheritance; Practical Breeding; Behavior; Diseases, Parasites, and Pests—Their Treatment and Control; Feeds and Feeding; Housing; Commercial Squab Production; and Pigeon Fancy—Exhibiting, Racing, etc. The first figure in the book is a photograph of a terracotta "Dove on Pomegranate," excavated from the Lydian city of Sardis (*ca.* 400 B.C.) and now in the New York Metropolitan Museum of Art, while the 906th one on page 612 shows a modern Texas pigeon loft complete with weather vane and a circular window made of a wagon wheel.

Understandably it took me a long time to settle down with this treasure, for I could not concentrate on any single topic with-

out being drawn away by others. The book still gives me the
same trouble. However, eventually I forced myself to pore over
certain sections of the genetics chapter, specifically those dealing with Fantails and Nuns. Here I ran at once into a kind of
disillusionment arising from the fact that almost everything I
proposed doing had already been done. For example, the mechanism of inheritance of Nuns' crests had been worked out so
clearly that the trait could now be listed as a recessive character,
designated "cr." And on page 330, Mr. Levi reproduced a photograph of a piebald pigeon with spade-shaped tail captioned as
the result of a cross between a black and white Nun cock and a
white Fantail hen!

Of course it is perfectly reputable to duplicate what other
people have done. In some departments we behave that way all
the time, going to movies recommended by reviewers and visiting the national parks or Caribbean islands our neighbors rave
about. But some of us unfortunate ones are adventurers who con-
trarily eschew those very movies and wouldn't be caught dead in
the touted parks or islands. Such a one am I, so I searched
through Mr. Levi's pages in actual fear of discovering the record
of a crested Fantail. The fact that I did not meet that further disappointment is not particularly remarkable. As I can now appreciate, domestic pigeons have been bred to possess such an array
of patterns, in both color and form, that many still remain to be
combined for the first time.

Things began to look brighter from a couple of other standpoints also. The Fantail x Nun cross had been carried only one
generation further, through a brother-sister mating, and then
dropped. The field was open beyond that. And as for the "fantail" itself, astonishingly enough this had not been worked out
genetically in full. I suddenly realized that my proposed new
pigeon variety would involve me as much with fantails as with
crests, for as soon as I produced a hybrid I would lose the full tail
and would then have to get it back in some way or other. If I
wanted adventure, I could have it without deviating from my
original plan.

Besides, it gradually became apparent to me that there
might be interesting behavioral traits as well to study from the ge-

netic standpoint. The Nun pigeons stood erect in conventional bird fashion, but Fantails executed several peculiar antics. Head bobbing was the result of neck shaking, a trait dubbed Zitterhals by German fanciers. Bending over backward, so that the head touches the root of the tail, does not seem to have a special name, but the medical term opisthotony, a sign of poisoning by strychnine, for example (from Greek *opistho,* behind; and *tonus,* stretching), fits it precisely. Dancing may or may not be a separate form of behavior; possibly it is only the result of the Fantail's trying to see what is going on while its head is in that awkwardly extended rear position. Finally, a good Fantail carries its tail elevated or cocked up—probably as a further consequence of opisthotony. These four factors, if such they can be called, had apparently not been traced through systematic breeding experiments, but perhaps I would be able to discern patterns in their individual inheritance.

It would be desirable to define clear-cut standards for each trait in order to measure them objectively from one generation to the next. In some cases that would be hard to effect. For example, with what frequency must a bird bob its head to be credited with Zitterhals? How far do you have to lean backward to become opisthotonous? Fortunately a crest promised to be either present or absent, according to the book, and I would not have to worry about intermediate degrees of that trait. As for the fantail, I found that the highest number of feathers ever recorded in a champion specimen was 48. My pair, for which I had paid Fred's friend only four dollars, could not have been of terribly high-class stock at that price, and their tails were modestly endowed, the male possessing 32 feathers and the female 30. Of all the Fantail progeny they had produced, I had counted the tail feathers of only their last three squabs, and these had had 30, 30, and 29. I therefore decided to set 30 feathers as the standard to which my birds must revert at the end of the experiment in order to qualify as having reconstituted themselves as pure Fantails.

These considerations brought up the question of the purity of both my beginning pairs. Was the female Fantail inferior because her tail had two less feathers than the male's? I noticed further that a couple of her upper tail coverts had very fine narrow terminal dark fringes, and some of her offspring had inherited the same trait. Was this evidence of some hidden taint, indicating that she was neither 100 percent white nor 100 percent Fantail? Yet her twenty sons and daughters had been indubitable white Fantails, so whatever impurity may have sullied their mother's past, it had now been safely snowed under with more recent immaculate genes.

When I had stood in that roomful of pigeons with Fred Kludzuweit, he had pointed to a couple of chocolate and white Nuns and said, "Those are children of the pair I gave you. They are not as good as the parents. That pair always produced squabs with a few misplaced chocolate feathers here and there."

So I was contending with the possibility of impure Nuns also. In reading what Wendell Levi had to say on the subject, I was shocked to learn that it is considered permissible to pluck out a wrong-colored feather or two from show birds in order to make them look right. (But not the reverse—you may not tie in alien feathers.) The standards were evidently not as rigid as I had thought they would be. Since that was the case, I would probably still be doing acceptable work if I began with fairly good Fantails and Nuns and produced the sort of crested Fantail that could be expected from such less-than-perfect origins.

My program was based on the assumption that if nature would cooperate by maintaining the constancy of her rules, everything should move forward smoothly. To be sure, Mendelian genetics is practically a mathematical science, and certainly you can depend on uniformity when it comes to basic arithmetic. However, my birds weren't exactly beads on an abacus, either, and there could easily be upsetting forces from standpoints far removed from the mechanisms of inheritance—everything from diseases to theft, for example.

Immediately, in fact, one heavily retarding influence impressed itself on the study. Whereas pigeons almost invariably lay two eggs to a clutch—and the female Nun had hatched a setting of two with her Nun husband—she now presented her new Fantail mate with only one. Even with twins pigeons reproduce relatively slowly, but now when I was all impatience, to cut the rate in half put a real drag on progress.

Nevertheless the inaugural egg had appeared remarkably promptly, so perhaps I should not be too critical. I had paired these birds on March 2; on March 4 they had displayed each other's acceptance by mutual preening; and the egg appeared on either March 9 or 10—I'm not sure which because I had not thought it necessary to begin checking carefully so soon. It may be relevant to observe that the female Fantail did not lay her first hybrid egg until March 25, a full two weeks later. Thus the speedy mating behavior of the female Nun may have arisen in a physiological necessity to accept any proximate male whatsoever, regardless of race, creed, or national origin. She must have been already becoming great with egg at the time that I separated her from her accustomed mate, and the process having advanced beyond a point where it could be arrested, her system in some fashion cried out for fertilization. This urge probably explains why even some happily mated females sometimes go astray and succumb to the ardent advances of desperately displaying strangers. It is comfortingly instructive to see the pigeon's mind and body so responsively interdependent.

As a matter of fact, I became quite devoted to that egg, simply because it *was* the first one—a material talisman for my intellectual gropings to rest upon. Yet I could not enshrine it for long. The downy squab that hatched on March 26 became hallowed at once instead, for a darkish spot on one side of its pink bill virtually shrieked "hybrid" when I delightedly observed it. As the little bird grew, I thought of another set of standards to watch in this and all future progeny. Fantails were white, of course, but Nuns had brown flight feathers, the outer eight primaries in the case of the male and nine of the female (another instance in which the birds were "impure," i.e., not exactly similar). Even while a squab's pinfeathers remain unsprouted in their sheaths, one can see whether or not each future feather is going to be pigmented. Thus I could easily keep track of every individually numbered primary feather on both right and left, to see if there was any regularity in the inheritance of color in those minute entities. But perhaps that was getting down to unnecessarily fine dissection.

In any event the squab grew rapidly. Within four weeks it stood before me fully feathered. But what trickery of nature was this? I have already confessed to a feeling of confidence in the laws of genetics. According to everything I had ever learned on that subject, I could expect any of three results after mixing a brown-and-white pigeon with an all-white one: I might get a further mixture of brown and white; or brown might be suppressed and I would get pure white; or brown might be diluted with white so that a new color, such as café au lait, would appear. But *black?*

The squab was pied, much like the hybrid illustrated by Levi. But the symmetrical pattern of the parental Nun had been lost, and I found it a bit mystifying that some pigmented feathers occurred in areas, such as the breast, where *both* parents were white. The head was white with scattered dark feathers, mostly on the left cheek. The six outer primaries on each side were black, two on the left having light tips. The tail was solid black. It contained 16 feathers. I should previously have remarked that the normal number is 12 (which is what both Nuns had), but how do you get 16 by crossing 12 and 32? Shouldn't the answer have been 22? I have mentioned the black spot on the beak; the squab's feet were similarly pied with black and pink. The crestless bird stood erect, without Zitterhals, and there you have its finished portrait.

Nature went on to manifest herself in both a constant and an inconstant manner, which I guess must eventually add up to inconstancy. What I mean is that the female Nun alternated between single and double clutches, a phenomenon that does not receive mention in Levi's exhaustive pages. Having laid two eggs with her Nun husband in November 1968, she thereafter produced the following with the Fantail during 1969: one in March, two in April, one in May, two in June and one in July. Quite unexpectedly she died on August 30, 1969, terminating her unique record. I now blame myself severely for having considered myself too busy that day to perform an autopsy on the carcass. Possibly the Nun's reproductive tract would have disclosed some interesting or revealing anomaly.

All those eggs hatched, but one squab failed to grow up due to its falling out of the nest and dying of neglect before I discovered it. I can summarize the condition of the six survivors as follows: *All* were pied black and white, without a trace of brown.

Minor correction: There was one whose tail was a *dark gray*, with a terminal black band, but gray is a dilution of black and white, not brown and white, so that I still could see no brown in any of those birds. One tail consisted of 17 feathers, two of 16, two of 15 and one of only 14 (average 15.5). "Fantail," as a genetic trait, thus seemed to be ephemeral indeed. It acted as if prone to being bred out in scarcely more than a generation. How, then, could I talk so glibly about my birds' reverting to that ancestral type?

My thought in carrying on the experiment in duplicate had been that I would thus be more sure of achieving a positive result—or at least results of some sort—through one of the lines. If the female Nun had died right away, rather than leaving six fine black-and-white hybrid descendants, I would have blessed that reasoning. Certainly the one outcome of such precaution that had not occurred to me was the possibility of gaining two sets of results at variance with one another, but the fact is that the cross between male Nun and female Fantail resulted in a race of *browns* and whites, much more the sort that I had expected in the first instance.

In order to make the comparison as fair as possible, I would like to single out the first six surviving progeny of this union, describing the first one in full and then presenting counts and averages for their tail feathers. However, I am unable to do this, for these offspring actually fell into two categories, one type of which was wholly or almost all white, the other showing variable brown patterning. It thus so happens that the first squab (whose nestmate egg was infertile) grew up into a pure white bird. The second will therefore have to suit my purpose. Its face, forehead, and chin were chocolate. The body was principally white except for a chocolate back from between the shoulders to the tail, and one black upper tail covert (that hidden ancestor still trying to assert himself!). Tail feathers were chocolate, but all primary flight feathers white. Bill brown and feet red, not pink. No crest. Posture and behavior normal.

The first six birds had 22, 18, 17, 16, 16, and 16 tail feathers, averaging 17.5. Since the mother had only 30, I would here have expected the average to be 21 (instead of 22 as in the other line), so it looked as if chances for eventually regaining the fantail might not be as bleak on this side of the composition-board partition as on the other. Indeed that one squab with 22 feathers was actually over the required mark, showing that I could count on some random variability along the way, rather than being bound inexorably by preordained genetic quotas. (You see, I was willing to let mathematics become careless when it worked on my side.) Yet this pair later produced some descendants with only 14 tail feathers, equal to the low score of their sequestered neighbors, so I did not let myself get too elated.

The greatest genetic inconstancy of all appeared in one squab of the third set of eggs. When this baby was half-grown, I noticed that the feathers of its head seemed very untidy. Soon afterward I concluded incredulously that the bird was developing a crest, or at least an approach to the crested condition. According to Mendelian rules such a character should be displayed on an all or none basis, and that is what this one finally did: The splendid chocolate-and-white female that eventually flew from the nest had a perfectly formed crest like her father's.

But that, of course, broke the other Mendelian rule which said that she, having a crestless mother, could only carry her father's trait unseen, because a recessive trait needs a double dose of genes to express itself. This situation can be easily diagrammed as follows, if the crested condition be denoted by double plus signs, crestless noncarrier by double minuses, and the crestless carrier state by minus-plus. Thus:

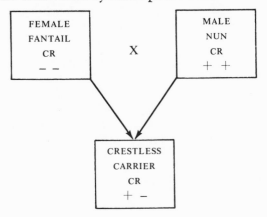

During reproduction each parent can give only one of its genes to an offspring. Thus if one crosses two crestless carriers (as I proposed soon to do), the possible progeny would turn out thus:

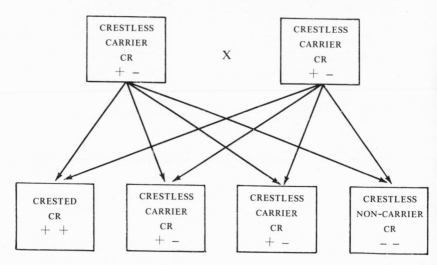

Hence one can secure crested grandchildren at a rate of 25 percent of their generation.

How, then, had that singular first generation bird turned up with a crest? Answer have I none. Nature was definitely off her prescribed track. Perhaps that devil had been in there, tampering with the works. It is too bad that the female Nun died, for if she had had more progeny, perhaps a greater variability would have appeared among them. Those six were in general so uniform, however, that I doubt it. To me it had seemed from the outset that my female Fantail may not have been everything that one might desire in a purebred bird. While I cannot put the full blame on her with any assurance, I am nevertheless inclined to exonerate her Nun swain. But whatever the truth, this was certainly a more entertaining line than the first one. On the other hand, I now had all the more reason to keep both lines going, for enjoyable though I might find freaks and monkey wrenches, I must still adhere to the possibility that a more sober path might be the one leading to my desired end (which is not to say that the more rollicking one might not get there, too).

It wouldn't do to recognize red herrings. The next step, obviously, was to sponsor brother-sister matings, each within its own line. This was a most tedious step—in fact it was more like a standstill for a long time. First of all the birds had to grow up and become sexually mature. Along the way they went through a post-juvenal molt when they all looked dreadful, like acne-faced teen-agers at the awkward stage. Then, when sex dawned on them, they did not know how to conduct themselves efficiently. In the beginning they laid numbers of infertile eggs, and later their squabs died of neglect. It was therefore almost a year before they began to produce vigorous offspring in steady numbers.

Now some new information trickled in. The black-and-white strain had children more variable than themselves. Some were entirely white, while others exhibited the "missing" grandparental chocolate. The tail feather count of one bird dropped to a chilling 12, but another squab had 18, more than either of its parents, allowing me to feel hopeful of some plasticity in that regard. In general it appeared that progeny of these brothers and sisters inherited tails of the same general quality as those of their parents, a logical enough consequence of these squabs' still deriving 50 percent fantail genes from their double grandfather. And here and there I reveled in seeing a crested grandchild.

In the adjoining family tree, matters progressed even more slowly. But meanwhile I stumbled on an unexpected clue that might clarify the whole project. After the female Nun had died, I felt sorry for the disconsolate male Fantail and gave him one of his nubile hybrid black-and-white daughters as a toy. Now I saw the advantage of experience, for while brothers and sisters were still faltering in the explorations of love, the Fantail rapidly taught his daughter all the rules and they began raising a succession of families at a rapid rate.

I had no interest in these children, for they were a step away, rather than toward, the crested state. In precise terms, their mother was a crestless carrier, the Fantail father was a crestless noncarrier, and consequently one unidentified half of the offspring would be crestless carriers while the other half would have entirely lost even the potentiality of the trait. Still, I found it amusing to see how some of the Fantail characteristics were strengthened in what could now be called a 75 percent Fantail–25 percent Nun breed. As might easily be predicted, pure white was the predominating pattern, though both black and brown showed up in small patches on occasional birds. But some of them had actual little fantails, thinly feathered to be sure, but spread in the fan shape and held in at least a partially cocked position. Feather counts of the first six squabs ran 26, 24, 23, 22, 21, and 18, an average of 22.3—a great improvement over mother's generation's 15.5. Moreover, a couple of grown squabs began to bob their heads, while one of those stuck out its chest as well while leaning over backward!

Curiosities? In a way that is all they were, but as I have implied, they were also my guide for the next maneuver in seeking crested Fantails. Brother-sister matings are commonly used by animal breeders, and that is the only genetic stratagem I had thought about, for some untutored reason. But parent-child combinations—so-called backcrosses—are also widely practiced, for utilitarian reasons rather than amusement, as in my case thus far. In contemplating the results of this particular backcross, I suddenly realized that I could use that method for cultivating the fantail trait in combination with the more strictly Mendelian formula for conserving the crest. I must simply use them alternately generation after generation.

Here is how it would work. For maximum clarity I shall begin at the beginning and number each step.

1. Cross Fantail with Nun. These birds are in what geneticists call the parental generation, and the cross is designated simply as P x P.

2. Progeny of P x P are the first filial or F_1 generation. In this case they are all crestless but are nevertheless all carriers of the crest trait.

3. Crossing F_1 x F_1 will yield the F_2 generation. Seventy-five percent of the individuals will be crestless; among these one-third will be noncarriers and two-thirds will be carriers like their F_1 parents. All these can be discarded. But the remaining 25 percent will be crested, and they become our jewels or chess pieces or what you will.

As I have already opined, birds of this generation can be regarded also as possessing 50 percent fantail genes, from the standpoint of their tails. The trick now will be that I shall treat the crest as a qualitative trait and the fantail as a quantitative one.

4. Select *crested* F_2 individuals and breed them back with their Fantail *grand*parents. The F_3 generation will then contain 75 percent fantail genes, but still *all* be crestless carriers.

5. Cross F_3 x F_3 to obtain F_4 individuals, of which one quarter will be crested birds with 75 percent fantail genes.

6. Breed *crested* F_4's back with fantail *great great* grandparents. The F_5 generation will then contain 87.5 percent fantail genes, but again *all* be crestless crest-carriers.

7. Cross F_5 x F_5 to obtain F_6 individuals, of which one quarter will be crested birds with 87.5 percent fantail genes.

8. Repeat this procedure as long as necessary to obtain crested individuals with the full number of feathers in the fantail. They should by this time also be dancing, opisthotonous birds, uniformly afflicted with Zitterhals. Probably they will have reverted to pure white. At whatever stage they seem to be sufficiently endowed Fantails—perhaps already at F_6, but more likely at F_8 or F_{10}—the great great great, etc. ancestor Fantails can be restored to each other and the crested birds can be put together to perpetuate themselves in a pure crested Fantail line.

Thus far I have completed Step Number 3, but the devil assures me that we won't quit until we have twisted nature to our will. As soon as I have enough perfect birds, I shall send a pair each to Fred Kludzuweit and Wendell Levi.

Chapter 5

THE LAST DAYS OF
POLYPHEMUS

It is perfectly all right for Polyphemus silk moths to be double-brooded in Cape May County, New Jersey. Official moth books give permission for these creatures to produce two generations during a single summer at that latitude. My argument, though, is that they ought to be less reckless in the way they do it. If emergence gets started early in spring, and succeeding stages then go full speed through both cycles, a second lot of caterpillars should spin cocoons reasonably soon before the onset of cold weather. But let some of the insects dally, and the risk of failure becomes terrifying.

Take Poly 106–1969, for example. *Her* mother had been Poly 185–1969, who in turn was the daughter of Poly 76–1968. That 1968 grandmother had been a no-nonsense sort of moth which emerged conventionally on June 20—neither early or late—and then mated dutifully with one of her brothers when I introduced him into her cubicle. She forthwith laid 414 eggs and died with a solitary unlaid egg remaining in her abdomen. I liberated most of the eggs by tying papers on which the moth had fastened them to twigs of various trees on my farm; but twenty of them I confined in a protective net on a willow oak near the beehive at the woods' edge of my orchard. These hatched into similarly no-nonsense caterpillars which spun cocoons during the period between August 17 and September 23, obviously too late in the season for thoughts of a second brood. (The books agree to this plan also, stating simply that Polyphemus *may* be double-brooded in South Jersey.)

One of those cocoons yielded Poly 185–1969 on the following May 30. Though this was not an excessively early date, it was still a good three weeks sooner than her mother's debut. Something was wrong with the moth's wings—they never expanded properly and she could not fly. However, she was normal in other respects, for when I tied a string around her waist and tethered her to a catalpa branch overnight, she secreted mating lure effectively enough to attract a fine mate. On the following night she laid 227 eggs in a paper bag, and I managed to rear 70 of the ensuing caterpillars on my long-suffering beehive oak. What with their early start, the caterpillars began to spin as soon as July 22. With a significant portion of July and all of torrid August still ahead, the prospect of a second brood then looked sensible.

Polly 106–1969 was not the first of her siblings to emerge, though she was not very far behind either. A brother was the most precocious, appearing on August 12, and only three days later Poly 106 arrived. She had normal wings (as had the rest of the brood), so their mother's crumpled pennants had not been a heritable trait. Poly 106–1969 duly mated with another brother and presented me with over a hundred eggs before I set her free to distribute the rest of her load in the natural way. But even before this new gift of eggs hatched, I began to have fears for them.

August *is* torrid in Cape May County, to the enrichment of the resort economy. Nevertheless, after the middle of the month there is an ominous change in the quality of the nights. The air loses its July haze and stars shine with a forgotten brilliance. A Canadian atmosphere moves in from the northwest, still bland enough to go with summer, but lowering temperatures in the small hours of morning so that our sleep becomes more refreshing. That is fine for us pajama-clad homeothermic monsters, but what about naked little cold-blooded caterpillars?

Such feeders and breathers, whether large or small, can run only as fast as the mercury allows. At noon on a day in late

August, caterpillars chew away at leaves of their food plant as destructively as they do during any other favorable period of their growth. But whereas earlier in the season they would continue voracious defoliation at night with only slightly diminished speed, certainly not bothering about the unneeded luxury of slumbering, they now are brought almost to a standstill for some of the dark hours. Consequently their day-to-day increase in length and thickness must be retarded.

Actually the eggs I harvested from Poly 106 were themselves held up in embryonic development by the change in climate. When the tiny caterpillars appeared on August 27, I realized that they faced the even bleaker prospect of September with, indeed, a part of October for their final growth and ultimate cocoon spinning. Going over my notes for the season's first generation, I found that Poly 185's most precocious caterpillar made it from hatching to cocoon in forty days. At that rate, I could expect Poly 106's fastest caterpillar to spin on October 6. But I knew very well that these second-brood caterpillars would have to be slower, so that even the most enthusiastic one was going to run a close race with the first killing frost (which might be expected around October 15). Then what about the average ones, and—worse—the conservatives? If they all died, wouldn't second-broodedness seem a harmful trait, and wouldn't nature have seen to it long ago that Polyphemus did not take such gambles in Cape May County, New Jersey? Or would frost not exert the effect I expected—did the caterpillars know some way to survive the cold themselves and to find necessary food after leaves had been frozen and stripped from trees?

Thus my nets were to be the scene of a natural drama that might be written with a tragic final act. Indeed this particular play, put on annually in my woods, could well have various endings in different years, so that what I was about to learn in 1969 could not be taken as the revelation of some fixed biological process. Perhaps that reflection was already a hint as to the answer—*sometimes* second-brooded Polyphemus got away with it, and that made it worthwhile for the species to try it every year. But now I was being guilty of anticipating the outcome on a hypothetical basis—something I could not prevent my eager imagination from doing. Obviously that got me nowhere, and what I really had to do was sweat it out day after day with the caterpillars.

A heavy mortality occurred immediately among the hatchlings of August 27. I had tied the egg-papers to twigs of the most succulent oak I could find in my woods at that date, but nevertheless a large number of babies failed to take to the provided fare. Whether this had to do with the lateness of the season (when, it has been alleged, tannins may have reached unpalatable concentrations in oak leaves), I do not know. I had seen many newborn Polyphemus caterpillars refuse to eat either elm or red maple leaves (which are perfectly nutritious for them) at other times, and had selected oak now because that is supposed to be their favorite dish. However, about fifty larvae finally managed to worry their tiny jaws into the thick oak tissues. Once they succeeded in overcoming this initial reluctance or difficulty —whichever it had been—they fed faithfully and I lost very few after that (until spinning time) from any cause whatever.

Of course they *did* grow slowly. This meant that their total metabolism was retarded—on the average, that is, for I have already mentioned their responsive acceleration under a warm noonday sun. In the early mornings even their chewing was slow. Whereas a midsummer caterpillar would go chomp-chomp-chomp, these could manage only c-h-o-m-p . . . c-h-o-m-p . . . c-h-o-m-p. When it came time for them to molt, they would have to rest for several days while a larger skin formed beneath the old one, instead of rushing through the process in a matter of forty-eight hours or so. Digestion, respiration, excretion—all of these must have shared in the diminished scale of a performance. If I had been able to check the heartbeat of a caterpillar's dorsal abdominal blood vessel, I would undoubtedly have found it to match the sluggish rate of other functions.

As September wore on and the weather became ever cooler, I ran into a new—and unexpected—problem. Weeks before the advent of actual frost, some of the oak trees began to prepare for winter. Some of their leaves turned brilliant yellow; others simply withered to brown and fell away. I found that if I did not tend the nets daily, I might discover the caterpillars searching vainly for green food in a desert of inedible leaf shards. I had divided the larvae into five groups of about ten each in separate nets, for with increasing growth they stripped branches faster, and isolating them in small lots allowed me to disturb them less often when they needed to be transferred to fresh greenery. But that now became a critical feature that prompted additional wondering about nature's management of double-broodedness. I found that sometimes an oak tree which "turned" early did so totally, so that wild caterpillars on it would have had to come to the ground and gone on a trek through the woods for another tree still bearing green leaves. Since they had no image-forming eyes, this would have been a quest whose successful outcome would be a matter mainly of good luck. My captive caterpillars were surviving because I was solicitous. But how many of their peers were even now starving to death, while, ironically, there was still time for them to beat Jack Frost to their hibernacula?

That was an unanswerable question for more than the obvious reason of my inability to climb all the trees and find all the free caterpillars to count them. The woods might be supporting numerous Polyphemus populations, only some of which had the same birthday (August 27) as mine. Older broods were obviously in the more favorable position, while if there were any of later date, those would be in even greater jeopardy. But it was possible also that my protected larvae had by this time become almost unique. Double-brooded moths may well be plagued by double-brooded parasites, and exposed Polyphemus caterpillars may succumb to ichneumon wasps and tachinid flies more rapidly in September than in July. Natural forces of several kinds combine to keep Polyphemus in check from year to year. Young caterpillars are favorite food for birds. Probably heavy thundershowers, accompanied by strong wind, kill many hatchlings as well as older larvae during periods of their delicate molts. If these various factors were not at work, each when its adverse effect could be felt maximally, the woods should be overrun by Polyphemus caterpillars at summer's end. Yet there was not a tree showing defoliated branches except where I had been responsible for tying my nets. In my daily walks through the woods, I never met even one wild caterpillar.

As October 6 approached, I found increasing proof of dawdling among my larvae. Polyhemus apparently goes through five stages of growth, each one requiring a longer time than the preceding. Thus when I noted the first caterpillars in their fifth and final dress on September 29, I was able to conclude immediately that they would be in no condition to spin a mere week later. And when October 6 finally arrived, a few larvae were still in the quiescent changing phase at the end of their fourth stage: These laggards had fallen an entire molt behind their parents' summer standard.

Now it was not too soon for an early frost to strike, if the season chose to be vindictive. Each night that passed without wielding icy knives could be blessed as benignant. Indeed night followed night, after the passage of that significant October 6, with a remarkably constant mildness for over a week. The days were those of Indian Summer (except for frost's shy refusal to come out of the wings) and the caterpillars made the most of them. On the fourteenth I noted that a few of the most advanced larvae finally looked big enough to spin *small* cocoons. This appearance consisted principally of plumpness, the caterpillars being rather shortened and hunched up so that their girth increased correspondingly. Nevertheless it was more than simply a matter of posture, for Polyphemus do not assume this pose, regardless of their bulk, until they have matured to some extent in the final larval stage; and before that time they do not have the option of spinning. But unless something dire happens, such as a failure of the food supply, the caterpillars don't dream of quitting yet. Their cuticle, though essentially nonstretchable, is still far from taut, and another week or so of chomping remains to be enjoyed.

That night it turned cold. At seven o'clock on the morning of October 15 the temperature was 41 degrees at the farm. There was no frost, but caterpillars were at a standstill in the nets. At that rate they were making zero progress for the first time, but of course within a couple of hours the air warmed up a bit and larval jaws resumed chewing, albeit in a halting, ponderous fashion. The following night was equally cold, although likewise without frost. Yet on October 16 I found a number of caterpillars spinning cocoons. These were now fifty days old, compared to their pioneer uncle who had spun at forty, and they were really ridiculously small, despite their grown-up stance, to have abandoned gorging themselves. But I had to admit that they were prudent. Two successive nights of near-freezing were surely sufficient warning that the real thing was next in line. Why risk even one more day, when feeding was now at such a slow rate anyhow that it would not contribute appreciably to further growth? Surely the wisest course was to spin as quickly as possible, if one had fortunately achieved spinning condition. For most of the brothers and sisters must still fight to gain that distinction—a fight in which the vital digestive processes were now pushing them forward at minimal speed.

This may then be the concealed gimmick that permits double-broodedness to occur in Cape May County: Some caterpillars squeeze in just enough feeding and growth to enable them to retire within cocoons and pupate at the very instant that the gong of death resounds. It is not for me to say that this is inefficient, even though that is what I happen to think, because nature probably knows what she is doing better than I do. But I feel that that these small cocoons may suffer a higher mortality during winter than those from single-brooded generations, simply because they *are* smaller and more fragile.

To go back to Poly 106–1969, the mother of these miniature October spinners—I had recorded the following statistics provided by her siblings. They had woven forty-three sturdy silken fortresses which (with their contained pupae) averaged 5.35 grams in weight. These ranged mainly from 3.9 to 6.7 grams, though there were two small ones at 3.5 grams and one giant at 7.7. Moths emerging from most of these would not impress anyone as being of abnormal size one way or the other, and there could be no question of their vigor. But what of this new batch of little cocoons? If these were normal or natural contingent in the Polyphemus population, I would expect to find a fair number of tiny wild Polyphemus moths in the spring; that, however, is something I am still waiting for. Perhaps they do not survive; or perhaps, as I have already suggested, they are very uncommon compared with the numbers I was able to rear in my nets. But in the latter case I must again question the utility and efficiency of the double-brooded maneuver, thus driving myself afresh into an unwanted position regarding the proprieties of criticism.

I deferred weighing the October cocoons until the entire crop was in and I could summarize them mathematically. Meanwhile the caterpillars not yet ready to spin must soon make hair-raising history. Some of them were bound to encounter frost in their nakedness, and I had no idea how they would react—whether they would be sensitive or resistant. On October 17 the radio for the first time made a long-range prediction of scattered frost for the night of the eighteenth in the Philadelphia suburbs. That might include South Jersey or it might not. In any event I was not going to be on hand to observe, since some unavoidable engagements called me away from the farm for the weekend. That was probably a good thing for me: I am suspicious of my own soundness when the affairs of caterpillars begin to impress me as hair-raising.

However, the prediction should have been of the short-range variety. I had planned to put on my store clothes, preparatory to leaving for Swarthmore after breakfast, but on looking out the bedroom window on the *morning* of the eighteenth I saw that fallen brown catalpa leaves on the lawn had all acquired a white rime overnight. Khakis and sneakers were obviously the ordered costume. The front porch thermometer registered 36 degrees. Running out to my closest forest net, I peered at some caterpillars to see what messages they would convey. Obviously it was not cold enough for them to be frozen solid, like jade carvings, though their immobility suggested such sculpture. And as for delivering messages, they had assumed sphinx-like postures, with their anterior portions elevated, as if in no mood to tell anything to a soul. At least they did not look frostbitten. Their green was as green as ever, and even the small red tubercle above each abdominal spiracle retained its seemingly useless gaudy fleck of gleaming silver.

I had time for one additional quick observation. Some of the most sensitive frost indicators are dahlias, tomatoes, and zucchini squash. All three grew in my garden. By the time I had finished breakfast the sun was shedding warmth everywhere except in the shade of the house and outbuildings. Remnants of frost remained in those shelters, but the garden was all in the open. I could not find a single blackened tender terminal leaf on any of my indicator plants. Thus I was forced to begin making excuses. This had not been a real frost. I was back to the original position of waiting for a valid ordeal with ice. Off I went on a not-so-carefree weekend.

Weather forecasters and alchemists must arise in the same ill-ordered womb, else they would not emerge so warped. Dawn in Swarthmore the next day came in at 50 degrees, while the maximum that afternoon rose to the 80s. On my return to South Jersey on the succeeding day it was still summer. Meanwhile a rash of new cocoons had appeared, ten, to be exact—over a quarter of the population. These, and the four that had preceded them, now looked rather silly alongside their still-feeding siblings. What had been an inferior fraction of the brood was presently gorging itself and rapidly catching up. If the weather should hold for several days, they would surpass their brethren who now appeared not as the prudent and wise but rather like fraidycats who hadn't dared to take a gamble.

On the other hand, these seemingly conservative early spinners may really not have reacted solely to the touch of cold weather. There are many triggers that set off various types of behavior, and it is sometimes a hidden force that produces results that we ascribe to more overt levers. Remember that these caterpillars were now considerably older than those which spun early cocoons a generation ago. Although the late-summer brood had done everything in slow motion, they had existed in the same sort of time continuum as their predecessors. Therefore, despite their laggard metabolism, they had passed through a greater number of day-and-night cycles. Somewhere in their nervous systems this phenomenon may have registered itself, and the consequent response may have been that a message permeated each caterpillar's body with the information, "I am an antique." Summer caterpillars would seldom be ridden by that kind of internal stimulus. Unless exhaustion of the food supply caused a famine, they would grow quickly until something inside them said, "I am huge." The urge to spin might thus arise from an assortment of conditions, operating alone or together according to seasonal circumstances.

Despite all the caution I used in seeking causes and ascribing motivation in these affairs, the caterpillars confounded me by responding individually. After all, if the hypotheses I had invoked to explain early spinners were valid, why did they not all hasten into cocoons? Why were many of them still feeding? How could they divide themselves into two groups, having all been born in the net at the same time and sharing exactly the same environment ever since? A comprehensive answer to these questions is that I was not playing with a chemistry set or dealing with matters of astronomy or mathematics. Each caterpillar was a biological system unto itself, varying from its siblings just as all of those, in turn, varied from one another. And in truth I did not even have two wholly discrete groups—early and late spinners respectively—but, as gradually became apparent, a gradient along which one caterpillar happened to be first, the others then stringing themselves out in an irregular pattern.

But not for long. On October 22, a remarkably late date for harvesting tomatoes, I picked a whole basket of beauties from vines as richly green-leaved as in summer. I noticed also a particular caterpillar in one of the nests that had set off in a restless search for an acceptable spinning site. During the night, frost finally got in its one-two punch. The twenty-third was bright, windy, and cold. Dahlias, zucchinis, and tomatoes curled up black. The caterpillar, which should by now have made a good beginning at enshrouding itself, remained stationary in the neck of the net, not yet having extruded its first strand of silk. One did not have to wonder if it were too cold—one could almost feel its numbed bewilderment.

Yet that had been only a touch of frost. The succeeding night brought mercury tumbling to 22 degrees and the axe fell in accompaniment. I was out before sunrise, in order to observe caterpillar behavior—or lack of it—before even weak rays of warmth could mitigate the situation. The larvae had all abandoned leafy twigs during the night and had congregated at the neck of each net, as if seeking to escape, though they could not have found warmer spots elsewhere. The restless one had fallen to the bottom of its net; obviously it was dead. One fat caterpillar had congealed into a solid icy stone, although another one, an inch away, remained soft. To the last minute they were maintaining their right of independent responsiveness! It now remained to learn whether a thawed caterpillar would return to life and whether near-frozen ones would survive longer.

Those answers came quickly. At midday, frozen caterpillars stayed dead, while living ones clung immobilized to their branches. Next day the few survivors chiefly remained motionless, though an occasional one managed to feed sparingly and—I am confident—ineffectively.

Very well: It would seem that the episode was complete, except for weighing the successful cocoons. Of course success could not be measured until next year, when the emergence of a moth would be the one acceptable critical measure. Meanwhile it was possible that currently cold weather would kill some caterpillars that had not yet completed pupation inside their hasty winter envelopes, and those cocoons would be immediate failures. However, for the present I was regarding all cocoons leniently, as being at least one step more successful than caterpillars caught nude by frost.

I now had twenty-one such October containers, flimsy though some of them were. Their statistics convey the clearest message when compared with the antecedent July generation.

	Heaviest	Average	Lightest
July cocoons	7.7 gm.	5.35 gm.	3.5 gm.
October cocoons	5.3 gm.	4.14 gm.	3.3 gm.

The average figure discloses that the stature of the second brood individuals was only 77.4 percent as great as that of the first—a decline equivalent to our six-footers' uniformly giving rise to progeny only about five feet tall, which would set us all screaming for vitamins and hormones.

Now it was time to decide what to think about all these doings. The last days of Polyphemus in South Jersey had been dismal ones, fraught with some outright failures, but also with triumphs that had dramatic merit even though the results were of meager quality. Would it not have been better to avoid the risks, especially since the goal could be only an inferior prize? Why did nature make the mistake of permitting some Polyphemus to enter the second-brood sweepstakes in Cape May County? Those individuals of the first brood which had *not* made such an ill-advised attempt were now slumbering, safely and fat, with far better prospects for a prosperous 1970 than their poorly grown nephews and nieces. Was it true that nature had actually erred?

I knew that I was not clever enough to decide better than nature. Therefore I sat down to puzzle out the whole proposition. Unquestionably it was in the matter of interpretation that I was seeing failure where failure did not indeed exist. I went back to my books and reread the part which said that Polyphemus ranges over most of temperate North America, being double-brooded in the southern part of its range. And there, both within and between the lines, was the answer!

The trouble had lain with my provincial insistence on judging the performance of my particular Polyphemus pets as if the welfare of the entire species depended on what these individuals did in Cape May County. But if the total continental population of Polyphemus was broadly divided into single- and double-brooded contingents, there must be a zone where one group merged into the other, and along that swath there might very well be a bit of trouble. A parochial view—such as mine—would then detect what looked like crossed wires, bad planning, or other thoughtlessness on the part of nature.

But I came to look at it as follows: If, because of constant inefficiency (with occasional flat-footed failure) in this intermediate climatic band, Polyphemus were threatened as a species—and that means the threat of extinction—nature would long ago have adjusted matters so as to eliminate the root of the conflict, i.e., Polyphemus would have settled on one method to the exclusion of the other. What then? If the decision had been in favor of double-broodedness, the northern population would be sacrificed, for all summer cocoons would proceed at once to yield moths whose caterpillar progeny would be wiped out as they were preparing for winter. If single-broodedness had been elected, the southern population would be deprived of half its annual crop, and a single generation would have to bear the entire weight of predation and parasitization during all its stages. Either loss, whether suffered by the northern or the southern contingent, would be far greater than the minor and easily afforded sacrifice presently being made at the mixed-up junction of single- and double-brooded populations.

If I happened to live somewhere else, I wouldn't have gotten into this mess. Anyhow, when I have my next confrontation with nature, I shall try to project my mind beyond South Jersey's ubiquitous oak trees.

Chapter **6**

TURTLES, PLEASE

On one of my local birdbanding expeditions in Cape May County, I set up a couple of large-mesh Japanese mist nets alongside some salt-marsh pools just west of Stone Harbor, in hopes of catching sandpipers. A small graveled road branching off at right angles from the causeway gave me means for getting away from traffic and also from nosy people who might be curious—or even resentful—about what I was doing. Sometimes these marshes, and the pools as well, are entirely dry, but winds, tides, seasons, and the moon play games with each other in that matter, so that at other times the whole terrain lies under salt water overflowing from Hereford Inlet. On the day I am now thinking of there was a stalemate in that tug of war, the intermediate condition allowing a few inches of water to lie in the pools, while I was able to squelch through the marsh grass on a merely muddy footing.

If any one had asked me to name the native creature least likely to be caught in my nets, I suppose I would have said something facetious like "a box turtle," just as I would not expect to find a shark flopping at my back door on the farm. But if I had not intercepted it in time, I would most certainly have had to untangle a box turtle from the lowest trammel of one net stretched across a marsh pool. What this terrestrial species was doing in a habitat that was practically marine, who can say? It looked perfectly normal—well nourished, uninjured, possessed of two eyes, four legs, and so on—yet it was assuredly in an abnormal setting.

I conjectured that some human being might have transported it to Stone Harbor. Now it had escaped and was plodding through the tidal pool (yes, it *was* headed west) toward the mainland three or four miles away on *my* horizon and simply at some unseen distant point that-a-way at turtle-eye level.

A number of weeks later I happened to be at the same place and thought for a few moments that perhaps I had found the actual human being responsible for translocating that box turtle from its natal woods and meadows. However, I soon realized that this eight-year-old red-headed boy would have been much too conscientious and devoted to let any turtle of his get away. I saw him and a middle-aged woman approaching from the end of the graveled road, where it meets a channel, and felt mildly annoyed at the presence of other people on "my" banding grounds. But the boy's first words totally disarmed me.

"Have you seen any live, large turtles?" he asked.

Don't think for a moment that this was an odd question. We turtle lovers get right to the point. A "no" or a "yes" promptly settles the matter and we adjust to whichever alternative is forthcoming, in order to get on with what we must do next. That question, by the way, usually draws a "No," as I now reluctantly had to concede to the boy. I told him of my recent experience with the seagoing box turtle, much to his titillation.

Recognizing a fellow indoctrinee, despite a small difference of five decades between our ages, I sought to give him something pleasant to remember by showing him a Willet's nest beneath a clump of grass alongside the road practically where we were standing. The nest contained four large pointed greenish eggs with blackish splotches, and the boy's hand went out at once to pick one up.

"No," I said. "I'm sorry, but Willets are protected birds and nobody is allowed to take their eggs."

"Not even the *police?*" asked the boy.

His aunt then showed me a tiny diamondback turtle she was holding in one hand. She said she had found it in a shallow hole at the end of the road—it had fallen in and was too small to get out. The boy, she said, shaking her head, was not interested in the baby, as he liked only *large* turtles. But she was bringing it back to their summer cottage for him anyhow.

There were three interesting points in her recitation. First, she was revealing herself as another human turtle toter, and it may actually be proved ultimately that this is one of the inborn traits of our race. Second, she had unknowingly informed me of the site of a diamondback turtle's nest; the little one in it had simply not yet managed to get out. Third, and by far of greatest interest to me, was the inexplicable preference this boy had for large turtles. Inexplicable, I call it, because I have always had an equally distinct but mysterious preference for small ones.

The closest I came in my young days to achieving knowledge and possession of a baby turtle, or—much better—turtle's egg, was when, as a school boy, I observed female eastern painted turtles digging holes in the sandy edges of a Massachusetts pond. I knew that they were about to lay, but by this time I was rabid for spring warblers and did not have time to sit and wait at turtle's pace for blessed events to occur. The females would have excavated their holes in varying degrees, from shallow saucer-like depressions just begun to cavities scooped out in globular shape at the limits to which the claws of their hind feet could reach to scrape. But my arrival always disturbed them and they would stop what they had been doing and plunge into the nearby water. In my hurry I would simply try to remember the locations of some of the females, so that I might dig out their eggs on my next visit to the pond. However, the turtles always packed the sand down and smoothed it over so that I could not find the exact spots. Some sort of wild creature was a better sleuth than I: The edge of the pond was often littered with leathery eggshells containing remnants of fresh yolk.

Thus when I began to explore the wildernesses of my farm, I approached the entire turtle universe with little more lore than a child could accrue. I knew that wild terrestrial turtles mate at several years of age, that the females then cover small clutches of eggs in shallow holes they have dug, and that baby turtles later clamber unprotected into a motherless world. My chief advantages now were that I had both the time and the authority to do what I liked about turtles. If a project should present itself, I would accept it immediately, no matter how much time it might threaten to occupy. Mentally I hung out a sign saying, "Turtles, please."

I have found nature study to be the most rewarding if you keep your approach as broad as possible—the wider your interests, the more fun you have. Now I would surely have languished if all I had cared about was turtles, because weeks would elapse between one turtle encounter and the next. But there was no occasion to count those days empty. Moths, birds, and mosquitoes kept my hours and minutes so full that turtles would actually have to displace them to gain their own audience. Moreover those sessions were invariably brief, consisting usually of no more than a formal greeting as we passed in the woods, but sometimes entailing a merciful euthanasia after a wanderer had been maimed on the road in front of my house.

All hail, then, the female eastern mud turtle that began hollowing out an egg chamber with her hind feet under a pitch pine growing on the open lawn near the edge of my woods on July 10, 1967! I almost passed her by, thinking she was nothing but a fallen pine cone—or really not thinking at all about the image that fell in my eye, for I was intent on changing a netful of luna caterpillars that had defoliated their persimmon branch. But via that peculiar reflex called the double take, my conscious brain suddenly translated the blob—the pine cone—into the clear vision of a small dirty-looking turtle in the classical egg-laying pose. I was already aware that eastern mud turtles were common in my so-called salt meadows which, being tidal, were moderately salty at some times but at others almost fresh and would perhaps best be called variably brackish to cover all the inconstancies. Mud turtles are slightly more marine than box tortoises, easily able to tolerate this mildly estuarine environment, judging by the numbers run over by cars and trucks where the road crossed my meadows. If I had depended only on the evidence of my walks in the woods, I would not have judged them to be that abundant. However, this kind of turtle does not often grow more than four inches long, and with its dark color and ungleaming shell, it would be inconspicuous unless it were in the middle of a cleared path.

Fortunately for me that pitch pine was neighbor to the persimmon on which I was rearing caterpillars, so that while I moved the net to a fresh branch and then transferred the fifty-odd caterpillars individually, I was able to glance at the turtle from time to time to see how she was getting along. In the usual way, she seemed to be doing nothing, and I began to fear that maybe I had drawn my conclusion too readily. Possibly she was sitting in that position by accident, not to lay eggs. As a matter of fact she had come an unusually long distance from the edge of the meadows —at least a thousand feet—through a woods whose littered floor presented fallen branches, logs, vines, hollows, and other snags in a continuous series. I had not previously found a mud turtle that far away from water. Of course this was a vigorous adult female, but what about her babies that would have to negotiate the same obstacle course in reverse? Could such midgets do it? Wasn't it likely that a mud turtle would serve her species better by digging holes somewhere near the meadows' edges? I suspect that some hidden force—that commonly named but never seen water table —might be responsible for the otherwise senseless trek. Surely the meadows' rim must lie at sea level, and at high tide there would be a tendency for seepage to extend inland unless a counterpressure prevented such movement. Eggs, laid even a few inches below the surface of the ground, might here be subject to drowning in the battle zone between saline and fresh media. My pitch pine was situated in what was—for this part of the universe —a relatively high and dry knoll, if only a foot or two higher than the meadows' margin. But whether a mud turtle could recognize that advantageous fact is something I'd rather let you answer.

When I had finished tending my caterpillars, some twenty minutes later, that dilatory little beast was still sitting there. I stooped down and looked into her eyes, but her stolid gaze told me nothing. However, I noted that the front part of her shell was crusted with a dried soil and sand mixture exactly matching the earth in which she sat, as if she had begun digging the hole head on (as I would have done) and then turned around when there was something going for her hind legs to work in. Now would have been an ideal time for me to play patience with her, at last being able to clock a turtle's least actions down to breathing and eye blinks. But my smiling mosquito pets had something to say about that. It is one thing to work actively at a job while *Aedes sollicitans* is tormenting you. You can slap away and hop from one leg to the other and change caterpillar nets or what-not. I assure you that I am unable to sit and watch inert turtles under the same onslaught. Regretfully I came away, resolving to return for a look every fifteen minutes.

The very first recess was the one during which she finished whatever it was she had come to do. I had found her at 9:35 A.M. Returning at 10:15, I saw her just beginning to walk slowly toward the woods. To make absolutely sure of her identification, I picked her up. Had she been a musk turtle (I refuse to use the alternative term "stinkpot," for no one has a right to call any animal a bad name), I would have known immediately by her malodorous conduct. I could have confirmed it further (though unnecessarily) by noting that her lower shell, or plastron, had only one hinge. This turtle remaining sweet and being double-hinged, I dubbed her "mud" at once. (Besides, I have still to meet my first musk turtle on the farm. Cape May County is within its range, but it is either rare or absent on my premises.)

I let the mud turtle go and turned to the spot she had vacated. This was still quite evident to me, but only because I knew where it was and what had been happening there. She had packed down the earth, oblitering the hole, and *that*, at length, disclosed that she had had more serious motives than merely loitering under an hospitable pitch pine tree. Yet the soil was sufficiently loose for me to dig into it cautiously with my two fore-fingers as probes. I tried to follow the course of the former cavity as closely as possible. The material dislodged was cool and moist as soon as I penetrated below the first half-an-inch. To my astonishment I did not find anything until I reached a depth of about three inches; it seemed almost impossible that the little turtle could have stretched her legs so far. But there they were, two fresh mud turtle eggs. In hopes that this would be but a beginning, rather than an end, I quickly refilled the hole, tamping down the sand to the same hardness I had found. Then I put up a circular wire-mesh enclosure at a radius of one foot around the spot, countersinking the barrier into the ground for an inch or so, both to keep out predators and to retain the hatchlings whenever they might come up into the world.

By September 2 it had been fifty-four days since the eggs were laid, or *three times as long* as it would take birds' eggs of similar size to hatch. So far as I could see, nothing had happened. The soil surface within the wire enclosure had not been disturbed, either by creatures digging from above or babies emerging from beneath—unless one of those events had taken place without leaving a trace. Possibly my disturbance in the very beginning had had some fatal effect on the eggs. Whatever the facts, I felt that I should possess them—which is saying that I could not restrain my curiosity any longer.

Over a foot of summer rains during the past weeks had packed the ground more firmly than it had been on July 10, so that one would wonder how interred eggs could survive and soft infants scrape their way upward in such a rigid medium. I could not use my fingers this time but had to resort to a garden spade. Immediately, however, I came upon the eggs, looking no different from their original condition. At the same time, since I had now moved a greater volume of soil than formerly, I uncovered three leathery old tattered turtle eggshells that looked not as if they had been plundered by a predator, nor yet as if they belonged to the same set as the unbroken eggs, but exactly as if they were remnants of a prior successful laying, from which hatchlings of another season or year had emerged. I could not tell that these were *mud* turtle eggshells; in fact they did look somewhat larger, though it was hard to be certain of that. But the discovery gave this nesting site under the pitch pine a dignity that goes with precedence. The mother I saw using it had not in any likelihood come to rest there randomly but was following a past habit of her own or a more generalized neighborhood turtle lying-in tradition.

The wisest thing to have done would be to bury the eggs once more and allow them to continue their interminable embryology—for I presumed that they were still alive because of their fresh appearance. They were not indented or shrunken, nor did any external discoloration suggest that they had rotted inside. But that would have meant a prolongation of boredom. Besides, I was beginning to face a couple of deadlines that made me want to speed things up now. Fall and winter were coming, and perhaps these July eggs were not destined by nature to hatch until next spring—a possibility that I could hardly believe. However, if the true course of mud turtles' life history in South Jersey demanded a very late fall hatching, I would be cheated out of witnessing it, for I was committed to leaving for Trinidad on October 16. Therefore whatever was to happen ought to take place as soon as possible, so far as my enlightenment was concerned.

I knew it would put the eggs in peril, but I decided to hurry their development in the house. Also I determined to keep them permanently in view, so that I would hear what they had to say the moment they said it. The greatest danger in this course lay not so much in overheating the embryos (for I would keep the eggs at normal indoor temperatures in the shade) as in dehydrating them. Their natural position underground would maintain them practically at the saturation point of humidity at all times, but their porous shells were not equipped with any mechanism to prevent water loss in even a moderately unsaturated atmosphere. My solution to these considerations and requirements was to place the eggs on paper towels over moist sand at the bottom of a loosely capped pickle jar on the sideboard in the dining room. There I could keep alert for the growth of mold if things became too wet, while avoiding the opposite extreme by sprinkling the sand instantly if it began to look thirsty.

Before depositing the eggs on their new bed in the pickle jar, I advised myself to give them a scientific going over, which is to say that I allowed my curiosity some further satisfaction. After brushing them lightly to clean off adherent sand, I lettered the eggs "A" and "B" in soft pencil so that I could henceforth recognize each one individually. Then I found out that they possessed the following vital statistics:

> A. 2.9 grams   22.2 x 13.2 millimeters
> B. 2.7 grams   20.2 x 14.0 millimeters

Moreover, I examined them against a bright light and found that both eggs were largely translucent, though each had a dark area at one side occupying about half the distance between the two ends. If those opacities represented embryos, they were certainly very far from mature. I could not detect any blood vessels coursing on the inside surface of the shells, nor did I see any movement of the dark interior masses. Thus the question of viability remained open, and my best hopes must still be based on the negative evidence of any decay having occurred.

By a strange chance, the August 25 issue of *Science*, arriving just at this time, contained an article on snapping turtles which gave the incubation period of that species as 55 to 65 days. Of course those eggs are large (five or six times as massive as mud turtles', as I learned later), so that one would expect them to be slow. But would that not prove immediately that something was wrong with my 54-day-old laggards? Now that I at last had some sort of standard for comparison, it seemed reasonable to me that mud turtles might take 40 or 45 days to hatch, but not longer.

I must watch those cloudy areas, to see if they increased in size. If embryos were indeed surviving, they must also begin to move, and blood vessels of the breathing membrances, pressed against the shell, should become visible. Alive or dead, the eggs were bound to lose weight. If they were dead, the loss would be due to simple evaporation and should follow a regular or linear course. On the other hand, living embryos would metabolize their yolk stores at an increasing rate as they grew, and change in weight would be further complicated by an uneven exchange of gases as lighter oxygen penetrated from the air and heavier carbon dioxide diffused outward from the biological engine.

Unquestionable changes did take place, right from the start, but they were of such minor degrees that I could not interpret them at first. When each egg had lost one-tenth of a gram after a week, was that a mere physical phenomenon or the result of vital metabolic processes? I found out on one occasion that the loss had been *my* fault, for there is such a thing as human fallibility and it is extremely easy to make different readings on two successive weighings of the same object even on the *same* day. Likewise I sometimes thought I saw a dark spot move or extend a leg, while the next time it would be as inert as an ancient corpse. That would not be an incompatible discrepancy, by the way, for even embryos have play times and then go to sleep, as every mother knows. When it comes to turtle embryos, the play may be so deliberately reptilian as to be hardly discerned, while sleep is of immense profundity. Therefore the criterion of activity was also difficult to apply.

The way things turned out, weight loss never became a revealing factor, no matter how smart I thought I was in outlining what that variation would disclose about the eggs' secret insides. And the first distinctly visible change was one that I had not foreseen at all—another point against my clairvoyance. On September 16, while examining them with my binocular magnifier and strong transmitted light, I saw an air space at one end of each egg. But of course! If either evaporation or transpiration were taking place, the amount of contained solid material must be diminishing and the volume it formerly occupied should become evident as a gap between the shell and the remaining contents. From that date onward things began to happen within egg "A." The light it transmitted became pinkish, and soon I felt certain of having identified both motion and blood vessels. Both eggs lost weight at a slow and constant rate, and in both of them the dark area appreciably increased in size, while their air spaces became more and more conspicuous. Yet egg "B" was enigmatic, refusing to give positive signals of being alive.

By late September adult turtles were again being killed in large numbers on the highways of Cape May County, including the road in front of my house. These were largely box turtles, though I noted some unidentified water turtles as well. I took this sudden increase in wandering as being possibly a prelude to hibernation, for frost was due in a couple of weeks. The reptiles may already have been looking for suitable places to burrow for the winter, since their cold-blooded nature must render them highly sensitive to the increasingly chilly nights at this season. Now I, too, was nearly ready to absent myself until spring, and I found it increasingly curious that my turtle eggs did not hatch with a rush so that babies would have a last minute chance to get a bite or two into their stomachs before having to go right back to sleep. Could they survive supperless?

On September 29 my worried attitude received support from a welcome but highly unexpected visitor. I found a newly emerged baby snapping turtle, umbilical cord still protruding from the center of its plastron, trying to climb through the wire into my pigeons' flying enclosure (possibly attracted by their bathing dish). It weighed 6.7 grams, which is exactly what I would have calculated for an unfed hatchling. So once again snapping turtles were rising to comfort and encourage me. There were scarcely more than two weeks left before I must go, but maybe my eggs would make it. Indeed I was now thinking seriously of taking them with me to Trinidad if they had not hatched by the sixteenth. Probably customs regulations would prohibit such traffic, but if I carried the eggs in my pocket, no one need know. For that matter, wouldn't it be entertaining—and surely a "first" for mud turtles—if they hatched *en route* as we soared 30,000 feet over the Atlantic?

I was supported in my thoughts of smuggling when we had our first frost on October 6. If indoor temperatures had not brought embryology to an end by that time, it was even less likely that eggs buried underground would have reached the hatching point. Thus it appeared to be likely after all—and against everything I would have predicted—that mud turtles lay eggs in July with the expectation that birth be delayed until the following spring. In that case I had undoubtedly upset the schedule by trying to force an earlier debut in the house. If I were to rebury the eggs on the day of my departure, they might be in too advanced a state to revert to a condition of suspended development. Possibly they would soon be torn open by fully formed babies on an Indian Summer's day, the turtlets optimistically popping above ground for a glimpse of the world just before winter jumped to slam icy doors on the soil and turn the babies to exposed frozen mummies. No, I could not have that: I must certainly break human laws as a penalty for having already broken natural ones.

On the eve of my journey, October 15, I weighed the eggs before lunch, to get a final reading for my records. "A" had lost half a gram in all, checking in now at 2.4 grams, while "B" had diminished by only three-tenths of a gram, thereby tipping the balance at the same figure. Then, at 3:40 P.M., I found egg "A" with a rent in one end and the tip of a living turtle's beak protruding! I immediately abandoned assembling my papers, packing clothes and so forth, in order to follow the next fabulous

historic moments. In characteristic chelonian fashion, nothing happened, as if the baby felt that it had done quite sufficient work for
its first day. Well, it was not enough for *me*. By this time I was
becoming weary of turtles' delays at every step of their lives. After waiting only half an hour or so, I recklessly peeled the baby
out of its shell.

Now I looked upon the smallest turtle I had ever *not* imagined, for my surmise had not shrunken down that far. In truth the
baby may not yet have been completely formed, because it still
bore a large umbilical mass of tissue. However, this appendage
did not bleed, and I felt that it was not something that must still
be absorbed into the abdomen. It looked, rather, like part of the
expendable extraembryonic paraphernalia that would ultimately
be sloughed away as an afterbirth. The infant's mind at least was
fully alert, for it was able to attempt rapid movements and could
right itself after I placed it on its back. On the other hand, that
umbilical nub was a bothersome handicap when the baby tried to
walk. In a pan of water it swam without much balance; unless I
had given it one of my fingers to climb on, I think it would have
drowned. Incidentally, I could see no trace of an egg tooth, such
as hatching birds possess at the tips of their upper mandibles for
chiseling their way out of hard-shelled prisons.

Meanwhile egg "B" lay inert. There was no occasion or excuse for me to tamper with it, for I could readily conceive of turtle eggs in the same clutch ending up with a week's spread—or
more—in their hatching dates. If I opened "B" now, provided it
was alive at all, I might find too great a degree of prematurity
to withstand my caesarian attentions. But my whole plan for traveling with an emergency obstetrical bag at the ready was now
changed. In the first place, I had accomplished much of what I
wanted to do: I had been present at the birth and I had seen the
baby. Secondly, I felt that the infant with its adherent cord might
not take well to the jostlings of a long trip. It would be piteous as
well as anticlimactic were I to arrive in Trinidad with no more
than an expired miniature companion. I dug a hole at the natal
spot, dropped in youngster "A" and egg "B," pressed down the
soil and set out for the Caribbean with the resolve not to think of
winter doings below ground in Eldora, but with doubts that I

could really put the drama out of my mind.

It was bright moonlight when I returned to the farm in mid-March—on the fifteenth, to be exact. By that illumination I could see immediately that crocuses were blooming in a row where I had planted them under a section of Rudd's old sheep fence that I had reassigned to duty as a trellis for young grape vines. Good! I had neatly disposed of the nongrowing season while sheltering under tropical palm trees, and now—despite the passage of five months—it seemed no more than a twinkling between the last time I had used the spade and the present shovelful that I turned. Not to say that I exhumed my treasures by the light of the moon at that very first available instant: In fact I felt quite proud of my restraint in waiting until the twenty-first, a full six days later. But then it was suddenly warm, spring peepers were piping in woodland pools everywhere, and I feared that the baby turtle or turtles, no longer enclosed by protective wire, might emerge and escape without my knowledge.

I had to search twice through the mass of wet soil before finding anything. There was no movement of any of the clods or lumps, and I had already forgotten the small size of the objects I was seeking. Even though my professional work had familiarized me with the metric system, so that I could easily think in grams, it was still hard for me to realize that mud turtle "A" had weighed only a shade more than a dime at hatching—much less than a full eleven cents worth! But at length I came upon my trove: an inert sand-and-mud covered blob of a turtlet and a fragmemted empty shell. To this day I don't positively know whether these were turtle "A," with the contents of egg "B" fully disintegrated; or whether turtle "A" had disappeared while egg "B" hatched underground to provide me with the baby I now held. This one carried no trace of an umbilical stalk, such as a newborn infant might have retained. Accordingly I adopted the unverifiable conclusion that "A" was the survivor. After all, egg "B" had never really done anything exciting, and it would be quite in character for it to have gone to pieces in the battle of worms.

I set a wide enameled pan on the dining room table, with shallow water and a few stones to climb on, and gave it to the turtle for its domain. At first the baby was extremely torpid, a state indistinguishable from being moribund; I would have been unwilling to decide between those diagnoses, though I feared that the hatchling might really be near death, again as the result of my prior interference with its normal life cycle. It kept its eyes tightly shut and responded only feebly to being prodded. But about an hour later it began to look about and seemed to have become more alert. It is evidently quite a more elaborate process to awake from hibernation than simply to get up after a single night's sleep.

Then, in the same tempo, more than one day passed before the turtle felt that it was ready for breakfast. For that matter, I did not know what to feed it. Even if the books had told me something about adult mud turtles' diet, those items might not necessarily have been digestible for babies. I felt reasonably sure that small soft-bodied invertebrates would be appropriate, though it was possible that mud turtles are vegetarians. Anyhow, whatever I gave the infant would probably be different from what it would have selected in its natural underwater habitat. Therefore I might as well avoid going to excessive lengths in foraging esoteric aquatic tidbits for it, but instead adapt it to animal fare that I could pick up right around the backyard. What could be easier and more savory than earthworms? Robins were pulling them up all over the lawn; surely I could do it as well as they.

For once something that I figured out ahead of time worked. I mean that I *was* able to catch earthworms, albeit I was not so picturesque with my shovel as the red-breasted birds with their handsome yellow worm-extractors. But that was an empty accomplishment after all, for I soon found myself accumulating a uselessly large earthworm inventory which the baby turtle refused to touch. On March 26, when it had been awake for five days and ought by any standards of survival to begin feeling hungry, the turtle eyed moving live sections of earthworm that I had chopped into bite-size lengths as if they were so many pieces of mobile but inedible detritus.

That discouraging though remarkable situation persisted for a whole month. Yet the turtle looked healthy and it even became more lively as the weeks went by. I now believe that it must have been living on yolk remaining when its yolk sac was drawn into the abdomen prior to hatching. During this period I changed the water in the pan several times, always adding some fresh sediment and rotted leaves from a pool in the woods. I could see plenty of captured wildlife scurrying about, so that the turtle could have chosen among a widely varied fauna. In addition I offered bits of fruit, such as a slice of fresh pear, just in case mud turtles really were—or started out as—vegetarians. I spent periods of every day watching everything closely, with my head bent over the pan, and if the turtle had shown even slight interest, I think I would have detected that expression on its blank face. But it was the old story again of the overanxious parent and the noneating child.

On April 25 something suddenly clicked. I had changed the water as usual and then added a canful of detrius from one of the rich pools. Almost immediately the turtle grabbed and ate a chironomid midge larva, deep-red and wriggling. Then it stalked a mosquito pupa but ate something else that diverted it by moving on the bottom: That creature was tiny and I could not identify it. Anyhow, that proved that the turtle was carnivorous at least, and moreover that it was now ready to feed and *would* feed under the artificial conditions I imposed by confining it to that restricted arena. I hasten to point out that this was not the first red chironomid larva the turtle had seen, for the tin scoop had brought in a few every time. This larva stimulated the baby simply because its appetite had arrived at last, whereas former ones had been unable to trigger reflexes that were as yet imperfect.

Well, now it was surely time to let the baby go, as it would obviously be able to look after itself henceforth. But suddenly something clicked in *me,* too. Of course I had become increasingly fond of the little pet, much as it exasperated me by its recalcitrance. However, if it would eat from now on, perhaps it would no longer be troublesome but, rather, it might be able to teach me further about the ways of mud turtles. I decided to keep it at least for a while, to follow its growth as it added to my education. As of that date I established that its shell measured 20 millimeters lengthwise. There: I had begun a new study.

Within a few days the turtle became addicted to the newly

discovered business of eating and grew voracious. The tiny animalculae from the woodland pool struck me as rather insubstantial, so I decided to try earthworms again. Now it attacked them immediately, though it had to put up quite a fight on two counts. When it bit into a worm, that creature promptly secreted abundant mucus, so that the turtle's mouth become fouled with it. Leaves, sticks, and other debris became entangled in the gooey mess, making it difficult for the turtle to proceed with further biting. It simply held on and tried to rub away the rest of the worm with one foot after the other. When it did sometimes succeed in doing so, it would be almost ready to swallow the piece thus bitten off, except for getting rid of a few more adhering fragments of sticks and leaves. But in rubbing those away, it might inadvertently spit out the segment of worm also. Secondly, the worm when freshly attacked at once squirmed violently. I would have thought this might intimidate the turtle, but not at all. It simply held on tighter and rubbed harder with its front feet, giving the worm more cause to struggle, and the two were soon meshed in Laocoön combat.

After several lessons the turtle became accustomed or resigned to the mucus. It no longer seemed to object, or else it found the luscious reward of fleshy worms sufficient compensation for ingesting a bit of slime. For my part, I settled on earthworms as the exclusive provision offered because of their aforementioned convenience. By midsummer the baby was so conditioned to this victual, and to my delivery of it, that it would rear up out of the water and snap worms from the end of a forceps. It grew most satisfactorily. On September 30 it had almost doubled in length, from 20 to 37 millimeters, while its bulk had increased over four-fold, from 2.4 to 10.4 grams. And it was *fat*. How can an animal living in a shell become obese? The trick is possible, and the way you verify it is by poking. A thin turtle can pull in its legs, head, and tail neatly, but when you poke a fat one here, causing it to withdraw the stimulated part, it inevitably spills out somewhere else.

With the beginning of fall the turtle began trying to burrow under rocks in its pan. Temperatures remained summery in the dining room, so it must have received a hibernating signal from some other cue—undoubtedly the shortening days. Now I sensed once again what my honorable next step should be, but I was still unwilling to let the animal go. I prepared a hibernaculum for it out of the same piece of wire I had used for its incubator. Without much ado it accepted the suggestion and disappeared on October 6 into the soil that I had loosened to ease its digging. I piled some dead plant material on top as a coverlet and wished it pleasant dreams.

With great expectations of a fond reunion I exhumed the turtle on the following April 6. "Exhumed" is the right word, for it was dead. I imagine the sand of the hibernaculum was too dense and became saturated, causing the turtle to drown, though that explanation could be wrong. After all, one reads of cold-blooded vertebrates hibernating under mud of ponds, where they ought to drown, too. However, I have always wondered if that is really true, for one can read also that during hibernation respiration and other functions are slowed *almost* to zero, but not *totally* so. Thus I feel that the animals should hibernate in a stratum that may be ever so slightly aerated but does at least afford minimal contact with the atmosphere. Perhaps amphibians could survive under ponds, by virtue of their cutaneous respiration, but not reptiles with lungs only. Whatever the truth—and it seems that some reptiles winter under water despite my reasoning—I suppose I should have foreseen that I might not select a suitable site and that the turtle's instincts would have served it better than my nonreptilian choice, if only I had been generous enough to set it free.

This experience had provided excellent basic training for my major concern with turtles, which of course remained to possess and nourish one or more box turtle eggs and to rear the resulting offspring. Somehow it had gotten noised about the neighborhood that I was a bug man. Kenneth Godfrey, who lived on a nearby farm, was just about the right age—I judge he was eleven or twelve—to appreciate such a person, for bug men and preadolescent boys have identical attitudes about a great many things. The first time Kenneth arrived, he was holding a medium-sized—that is, about three-foot-long—black rat snake in each hand, having nevertheless been able to maneuver his bicycle at the same time while thus encumbered. I admired the snakes, which is all he wanted me to do, and off he went. On another occasion he had found a beautiful newly emerged female royal walnut moth which I happened to want very much, though he had not known about my desire. I practically took it away from him while pressing fifty cents into his palm, despite the likelihood that he would have been glad to give it to me.

It gradually dawned on me that Kenneth Godfrey might be just the person to set on a box turtle egg hunt. The next time he visited, I asked him what he knew about such things. Wow! He said he had sixty captive box turtles at home as of that very instant, and that he had more than once seen "wild" individuals laying eggs—had dug the eggs up, in fact, though they had invariably shriveled and never hatched. Did he think he could find some more? Yes. Would he deliver box turtle eggs to me for twenty-five cents apiece? Yes. Would he start right away? Yes, again.

A few days later I found a sand-filled jar waiting for me in a shady spot in the garage where we had agreed Kenneth should leave his contributions if I were not at home when he came. I rushed eagerly into the house with it and dumped the contents into the now idle enameled pan. Out rolled two long, almost cylindrical eggs with the following measurements:

A. 3.2 grams   25 x 14 millimeters
B. 3.2 grams   25 x 14 millimeters

If you will check back to my mud turtle eggs of two years ago, I think you will agree that the new ones might be their kissin' cousins. Certainly they did not have the mass and girth that I would expect in eggs of the relatively gigantic box turtle. Besides, here was a mere pair of eggs once more, whereas I had the feeling of having heard (from some forgotten source) that box turtles normally lay half a dozen or more at a time.

As Kenneth told me later, he had not tried to trick me—he simply had brought what he had been able to find and let me decide whether I wanted it or not. Thus I withheld my excitement before viewing his next offering, and that conservatism of wasted emotionalism proved wise when I beheld an opposite kind of spectacle: five eggs that were obviously *too big*. These averaged 14 grams in weight and measured 37 x 27 millimeters long and wide, respectively, or about six and one-half times as voluminous as mud turtle eggs! I concluded at a glance that they had been laid by a snapping turtle. The best way to prove that would be to hatch them out, but almost every one had been badly indented during the nest's excavation, and I felt pessimistic about their future. Kenneth's failure to rear box turtles' eggs in the past—if he had really found any—may have resulted from too rough a touch. I kept both his sets in my lab for the rest of the summer, but they followed his formula by shriveling up. So much for trying to place my burdens on a neighbor boy's shoulders.

Meanwhile I had been keeping my eye on turtles in my own precincts and had actually had an exciting day when I found a box turtle with its rear end partially backed into humus at the woods' edge near the beehive. However, this proved to have been only a resting animal. The weather was exceptionally hot and dry at the time, and the turtle's maneuver in partially burying itself had been no more than an effort to get into the shade. Since I was now highly alert to any clue whatsoever regarding turtle eggs, I became the victim of further frustration at my very door. It seemed suddenly as if turtles of one sort or another were making nests all over the place, and that moles or skunks or some other predators were finding them, for every few days I would find a new plundered clutch, with its devastated shells littering the ground like a paper chase.

It is only because of the tension produced by those circumstances that I can explain why I should have dug up a diamondback turtle's nest at Stone Harbor and brought the eggs home to hatch. Surely I had had no previous interest in diamondbacks. This is what students of animal behavior call an "intention movement." I was so driven by the urge to dig up a box turtle's nest that I enjoyed a sort of quasi-fulfillment by digging up a substitute. At the same time I felt conscience-stricken, because the diamondback, though still very common in the tidal channels and salt marshes along the South Jersey coast, might become rare or disappear with further real estate development *and* with continued slaughter by automobiles. During early summer when the females come out to make burrows in the sand, they seem invariably to want to do so on the other side of a highway, and their smashed bodies sometimes litter the roadsides more thickly than beer cans. Conservation-minded people in Avalon actually erected two stenciled signs that said, "Please watch out for our turtles." With that kind of commendable public sentiment growing, you don't go around lightly robbing turtle's nests.

I had returned to the sandy side road off the Stone Harbor causeway where I used to net sandpipers and where the red-headed boy's aunt had found a baby diamondback in its nest hole. Now I was studying a colony of Seaside Sparrows, still paying only passing heed to the turtle fauna except for an awareness that the adults sometimes trudged through the marshes and might possibly not disdain sparrows' eggs for breakfast or nestlings for lunch—though perhaps I wronged them greatly by even considering such a derogatory suspicion. At the end of the road was a convenient place to turn the car and here, just as I arrived, I found a large diamondback turtle with what I can only call a most guilty look on her face, as if I had caught her at something she now wanted desperately to hide. Obviously I was reading my own interpretation into the situation, for a stupid turtle can't look or act anything but stupid. Nevertheless, I felt so sure that she had just been up to something secret that I got out and scrutinized her sand tracks. Sure enough, they led to a blank spot and stopped there, precisely as if she had arrived out of a buoyant atmosphere. So what could I do but dig?

Close by I found a handy tin can of large size in which to place the eight eggs, after first making a bed of sand and then adding more up to the top. I kept looking up to see if anyone were watching, but I was alone at the center of a great flat horizon. The eggs used to worry me at night, for fear the loss of these eight babies might be just the number that ultimately nudged the diamondback species toward extinction instead of survival. A few weeks later my heart sank to whatever bottom hearts can find when I met a bulldozer on the little road. The real estate people had arrived at last. Good-bye to my Seaside Sparrow project! However, it turned out that the land had been bought by the World Wildlife Fund, and that a large area of local marshes had been transformed into a permanent sanctuary. On the very spot where I had turned the car around, a small research station, called the South Jersey Wetlands Institute, would be erected. The sparrows were safe after all. But the bulldozer had irreparably dozed the site of my robbed turtle nest (as well as numerous old Willets' homesteads), so that if the eggs had still been there they would all have been destroyed. Now I could sleep.

One nice thing about those eggs was their size. The diamondback is a moderately endowed turtle, not far above a box turtle's bulk, and indeed both those species could be regarded as about midway between mud turtles and snappers. Well, maybe that was true of their eggs as a corollary. Those of the diamondback weighed close to 9 grams apiece, whereas I calculated the half-way mark between my mud turtle eggs and Kenneth Godfrey's big set at a little over 8 grams. That gave all the support I wanted to my belief that the boy's unhatched damaged specimens had in actuality come from a snapper. And of course *that* led to the conclusion that I had yet to see my first box turtle egg—which, if one ever did appear under my gaze, would (I now felt sure) weigh in at about 7.5 grams.

But what could I do with a clutch of baby diamondbacks? I did not need any further nursery training, having already become certified in turtle pediatrics. One set of my grandchildren solved part of the problem. On a visit to the farm, their parents had closed the car windows tightly to prevent swarms of *Aedes sollicitans* from taking up residence inside, and two baby red-eared turtles that they were transporting became irreversibly cooked to death in the overheated sedan. I had shown them the diamondback eggs, and the next thing I knew I had pledged to make good the loss of the red-ears as soon as nature came up with home-grown replacements. Thus two babies, wrapped in tissue paper in a cardboard box, took the parcel post route to Clinton, New York. Three eggs failed to hatch, and two other babies died soon after hatching, so I was left after all with only a single problem child. Naturally the easiest way to handle such a situation was to sit quite still and do nothing.

The new dining room inhabitant went through practically the same stages as my former one—the noneating period, followed by earthworm indoctrination, for example. Yet it taught me a new fact or two besides. For example, I found that it got along very well in fresh water, despite being a saltwater creature, although its skin almost constantly was shedding in ragged strips, possibly as a consequence of the bland medium. But, alternatively, that may have been only because the turtle was insufficiently active and could never scrape its way over a sandy surface to abrade its skin in a natural manner. Oddly enough it would never climb on the rocks I provided in the pan but remained in the water at all times. If it really had had skin trouble because fresh water was alien to it, I would have expected it to try to dry out. Furthermore it behaved in a perfectly contented manner, exhibiting no sign of hyperactivity or exaggerated efforts to escape. So far as I could tell, nothing was wrong.

When fall arrived, it duplicated the "intention movements" of hibernation that I had noted in the mud turtle, acting as if it wanted to squeeze under the rocks or burrow through the bottom of the pan. My farm would be an abnormal dormitory for a diamondback turtle. To liberate one here at this season might be the same as condemning it instantly. Yet I could envisage no better future for it if I were now to toss it into the channel alongside its old destroyed nest—I could almost feel the hostile army of horrible creatures waiting there to snap it up. Frankly I had no idea whether baby diamondbacks commence an aquatic existence at birth or wait until they are larger to enter the habitat of sharks and such. Once again I had broken a natural cycle, and I simply did not know where to introduce that baby into the position it ought to have occupied at this particular stage.

Therefore I repeated the maneuver of doing nothing. Perhaps keeping a turtle up all winter would not be quite as fatal as the courses I had discarded. It may be that I imagined it, but I think this one became torpid in October and then perked up, having somehow adapted itself to the privation of slumber. Anyhow, that was a small problem compared with food. Well into the fall I continued finding earthworms without any trouble, but as soon as the weather became really cold their supply diminished alarmingly. By December I would be lucky to find even one after spading through a large volume of my prized compost heap,

or else digging next to the foundations of the garage where concrete underpinnings seemed to have some lingering attraction
that delayed annelids' deeper descent. And then finally a blank:
no more earthworms.

My daughter Valerie in Clinton reported that the two diamondbacks I had sent were thriving on dried turtle food that
she had found in a pet shop. (They, too, by the way, were constantly shedding their skins; that was therefore a trait of the species or of the freshwater environment.) Accordingly I bought
some of the same brand she had named, as well as another sort
I happened to see. Between them these preparations contained
shrimps, fish meal, meat meal, and fly larvae among several other
attractive ingredients, and I foresaw that my diamondback might
be hard to wean back to worms next spring. But not by any
means was that true. Apparently a turtle trained to eat living food
fails to recognize good things to eat if they are motionless. The
shrimps, fly larvae, etc., absorbed water, swelled up, became
moldly, and finally rotted away. If the turtle ever ate any, I did
not witness the act, and it surely must have taken in a minimal
quantity if it did—hardly enough to stay alive.

Like the mud turtle, the diamondback had become fat before the supposed time for hibernation, and it was now apparently wintering on its stored adipose deposits. That was undoubtedly one way to get along, but I did not like it. Every time
I looked at the turtle I felt mean, for it must have been hungry
and that was my fault. What live game *could* I find for it in midwinter? Finally I discovered in late January that the bagworm
cocoons that hung all over my evergreen shrubbery and multi-
flora rose hedge contained a variety of small items that moved
and were edible, even though their contribution to nutrition could
not have been more than slight. But the first time I dropped a few
tiny white maggots into the water, my diamondback's fervid response settled the food question for the rest of the winter. Now
instead of feeling like Fagin when I looked at the abject turtle,
I could at least *do* something that was slightly generous. Many of
the cocoons I dissected were empty. Some contained maggots of
various kinds and sizes, which may have included both predatory
and parasitic sorts. A few held large dead ichneumon wasps

which as larvae had feasted on the bagworms and later been consumed by maggots themselves. And a few—perhaps one in ten cocoons—harbored next summer's generation of bagworms in the strange manner of that class of pests: a mass of fertilized eggs occupying the shell of their long-dead mother's body. Unaccountably my turtle relished those eggs, though they were obviously as immobile as anything can be. But then, they had not been dehydrated, either! Thus the turtle survived the rest of the winter, though it was a signal relief to us both when at last I turned up the first earthworm of the year in the compost heap on March 10.

Spring brought up the box turtle question anew, but not until I had given it plenty of additional, cold-weather thought. Actually my thinking was more borrowed than original, for first I located a helpful, though short, passage about keeping box turtles in captivity in Roger Conant's *Field Guide to Reptiles and Amphibians,* and from there I was led to Archie Carr's to-me-hitherto-unknown *Handbook of Turtles*—a veritable bible for devotees like me. Conant said, "Most [box turtles] adapt themselves readily to captivity, requiring only a backyard or a box of dirt for digging and a shallow pan of water for an occasional soaking. They are omnivorous, and are fond of fruits, berries and raw hamburger. Many people feed them table scraps."

What could be more sane, then, than to keep a harem of turtles right under my nose, so that when nature took its course, I should be there to revel in all its manifestations? I decided that one male and four females ought to be enough to do the job. As for knowing a boy from a girl, you may remember Ogden Nash's words about the turtle living 'twixt plated decks which practically conceal its sex (wherefore he considered it clever of the turtle in such a fix to be so fertile). According to Archie Carr the males *do* have to be a bit clever about it, being almost thwarted by their anatomical peculiarities. However their lower shells, or plastrons, are shallowly concave, which helps somewhat in making connections. A rather superfluous further clue to the recognition of male box turtles is that they often have red eyes, compared to females' brown ones.

Before other springtime pursuits became too demanding, I asked Rudd to dig up a small plot of lawn, 6′ × 12′, near the asparagus bed, removing all roots of grass and other herbs that might impede a turtle's future efforts to excavate an egg chamber. Then I requested that he enclose the area with one-inch mesh chickenwire, countersunk into the ground for about six inches, and also roofed over above in order to exclude predators from all directions. Rudd suggested that half the top should be covered with a corrugated metal roofing sheet that he happened to have stored in the garage, and I agreed that it might be a good idea to give the turtles a choice between exposed and sheltered sites for their digging. He finished the job in a couple of days, and I then added a luxurious appurtenance in the form of a sunken plastic swimming basin, containing rocky stepping stones, at one end of the pen. After this small expenditure of labor, I found myself again at a dead end, helplessly dependent on turtles for further adventure.

I was planning to go through my notes, to see whether I had made any record of the earliest spring date on which box turtles had appeared in my woods, but before I got around to that bit of research, Kenneth Godfrey arrived on his bicycle with two fine specimens, a male and a female. I had been giving him excess pigeons from my flock, and I must have mentioned my new project to him on one of his visits to collect the latest squabs, though I had not enlisted his help as a turtle collector. But this was a good way to pay me back, as well as to keep the flow of pigeons active, and I was almost 100 percent pleased with his gift. The only flaws I found in the situation were both personal and inconsequential: I should like to have found the turtles myself, just to keep everything totally and tightly within my private emotional sphere; and I should have preferred to harbor turtles from my own woods. The latter point might be of some importance, because box turtles adhere to rather small individual home territories, rather than wandering randomly throughout their lives, and thus a captive turtle within sight of familiar ground might be less depressed than one imported from a distance. Such a female might make love and lay eggs with correspondingly greater felicity.

The date was April 20. During the next two months I continued to see no turtles in my woods, but Kenneth kept reappearing, or else he sent his younger brother or sister, with further awakened sleepers. I have no idea why my own denizens should have been so sluggish. Anyhow, I thus ended up with an entirely foreign population in my pen, and at last I had to refuse the turtle torrent from across the way.

The male was a handsome animal, so red-eyed as to look vicious (although he was actually as gentle as any of the females), and with particularly generous splashes of yellow marking his head and carapace. But in all the time I possessed him, I never once saw him take any interest in any of the four mates that I practically thrust upon him. Of course there were many hours during both day and night that I was not there, and he could easily have mated repeatedly with all the females in my absence. But you would think that when I was present he might once have made some sort of suggestive pass at one of them that would signify at least that he recognized her as a desirable turtle—or even as *another* turtle instead of a mere clod.

I must make fair allowance for the mosquitoes. *Aedes canadensis* soon discovered the turtle pen and moved in one by one from the adjacent woods until each turtle had its own buzzing cloud of attendants. These tormented the animals day and night. One way to escape the pests was to disappear in the swimming pool, and it may have become a matter of sex versus comfort, though sex of course is supposed to be the much stronger drive. Else the turtles half-buried themselves in the loose soil or crawled under a pile of dead grass in one corner of the pen. Of course mosquitoes could follow into the terrestrial retreats, and the turtles may have behaved as they did for entirely different reasons. Perhaps they were even oblivious to the mosquitoes. The trouble was, they seemed to be oblivious to sex, too.

Here is where Archie Carr came to the rescue. Amazingly enough, he reported that box turtles (as well as some other species) do not need to mate every year in order to lay fertile eggs. Indeed, he knew of one case in which a female produced young after four years of celibate captivity. Perhaps my male *was* homesick for the Godfrey's farm and *did* neglect his seraglio. For that was another of Archie Carr's revelations (Ogden Nash was not so far off the mark about the mating difficulties of turtles): He wrote that the male box turtle can only barely manage to achieve union with the female, and also that females are at times rather coy, so that an eager male would probably have been chasing around the pen constantly rather than lazing in the swimming pool as mine did.

The next problem was to know whether or not—and when and if—any of the females laid eggs. I was ready to assume blithely that they were all carrying stored spermatozoa from wild matings in former years, but I did not know how to distinguish a pregnant one from an animal that had discharged its eggs, since obviously there can be no such thing as a pot-bellied turtle. If I failed to observe mating, I could just as easily happen to be elsewhere when the ladies took to digging holes, and because they always cover the eggs and carefully smooth the ground, I would again remain uninformed. Possibly I ought to recruit Kenneth Godfrey again, for if I had fifty females instead of only four, my chances of catching one in the act of oviposition would be much better.

Now the expression "prenatal clinic" began to float in my mind; I could see queues of towel-draped primiparae lined up before a scale as a nurse checked off the poundage each one had gained since last visit. That was it! Though I might not be witness to egg-laying, I would nevertheless know it had taken place if a turtle that had been gaining or at least holding steady were suddenly to show a massive drop in weight. I must inaugurate my own prenatal clinic for turtles. But how often should I weigh them, and what would be a significant loss? Perhaps those questions would answer themselves after I got the system going.

By the time I had this brainstorm it was already May 8. What if box turtles were ready to lay as soon as they emerge from hibernation? Then they might already have hidden their treasures from my prying search and their future weights would mean nothing obstetrical. Again I took heart from Archie Carr: In my latitude, box turtles were not likely to lay until June. It was therefore more than possible that they had emerged from hibernation with empty stomachs and were even now eating for both themselves and a future generation as their eggs matured. Thus I ought to be able to chart gains within the next few weeks that would provide splendid standards against which to contrast egg-weight losses. I did not much like the thought of what must happen next—that I would be forced painstakingly to dig up 31,104 cubic inches of soil in that 6′ × 12′ pen in a quest for eggs lying three inches deep. But that would be better than bulldozing my whole woods, besides which I would *know* that I was within breathing distance of fulfillment.

"Eating for two" was another expression from the prenatal clinic, but as a matter of fact my turtles were not particularly enthusiastic eaters. Indeed they weren't enthusiastic about anything, unless it were to promenade at the end of the pen *toward the Godfrey's farm* as if really longing for home—they quite tamped the earth hard with their feet in that area. But they upheld Roger Conant's statement by eating *some* hamburger and *some* fruits and vegetables, such as grapes and cherry tomatoes (that I bought at great expense, considering preseason prices at this time). However, now that I had thought of the weight system of detection, I wanted them to gorge themselves. That was not being a good doctor, in view of the conventional admonition generally being exactly the opposite.

I labeled each female turtle with a gummed plastic letter on her back—B, C, D, and E ("A" was the male). On May 8, the first day of the clinic, these weighed respectively 375.1, 424.8, 506.3, and 432.4 grams. On that date I found that if I put the turtles on the pan of my balance upside down, they remained quiet long enough for me to adjust the counterweights and get an undisturbed reading. In ensuing weeks, when they had become tamer or bolder, or perhaps just brasher, they tried to right themselves immediately and I came near to slapping their disobedient bottoms. But this was only a minor nuisance and I should say by way of generalization that I ran a reasonably orderly clinic.

On the following Wednesday I reweighed the turtles in alphabetical order. "B" had gained 12.7 grams. This represented 3.4 percent of her initial weight and would be the equivalent of a 110-pound primipara's gaining 3¾ pounds. That was probably fine on a montly basis in human terms, and I judged it to be sensible from a turtle's standpoint as well, unless it represented only undigested gastrointestinal contents. Well, *next* week should solve that one, if the animal continued to gain, for she could not go on merely bloating her stomach and gut indefinitely.

"C" had gained only 5.1 grams. Still, she had moved forward. "D" was up 11.3 grams—another modest but substantial gain closely matching "B's" 's increase. However, with "E" the system broke down, for she had fallen back a full 12.4 grams, as much on the negative side as my best two mothers-to-be had shown on the positive. But wait! Did this mean that she had already laid eggs? Judging from what I knew of egg weights in other species, I had already concluded that box turtle's eggs should weigh around 10 grams apiece, and that would mean that "E" had laid only one. Archie Carr's data led me to expect more than that—at least two and usually more, up to six or seven. Hence I had to admit that "E's" loss in weight was inexplicable. In that case, did I have a right to accept the others' gains in the light that I wanted to interpret them? Regrettably not: I must wait until next Wednesday to reweigh *all* the animals before drawing even a first tiny conclusion about any of them.

Then once more Kenneth Godfrey scooped me. He held out his hand one late afternoon, extending toward me two oval objects that I recognized instinctively the moment I saw them. Yet I remained just as instinctively cautious and asked, "What are they?"

Kenneth is a champion among those born to be laconic. All he would answer was, "Box turtle," when I wanted to know how, where, and whatever all at once. Gradually I got the information from him, albeit in monosyllables and disyllables only, and certainly never in sentences. His mother had run over a turtle with the Godfreys' power mower and had "chopped its back open." "Happens all the time." The eggs became exposed by this rear caesarian approach and were in fact "hanging out." I could not learn whether there had been other eggs, whether any had been broken, or whether he had cleaned these up. Anyhow, here they were, quickly transferred from his hand to mine, and—so far as I could see—intact. Of course they were fully shelled—or leathered, or skinned—ready to be laid soon, and that must mean that they had already been fertilized, because no sperm could possibly penetrate that tough, thick membrane. Hence I could see no reason why these should not develop and eventually give rise to baby turtles just as well as naturally laid eggs.

Again this was not the way I would have chosen it to be. On the other hand, I was not about to give up my clinic because of an otherwise superlative windfall. I was now simply going toward my objective with two approaches, one through the premature baby department and the other on an outpatient basis. I quickly resurrected the pickle jar with its damp sand and paper-towel bedding and in short order gave the eggs a prominent place on the dining room cabinet where I should see them daily. Oh, but wait! I weighed them first: They registered at 10.5 and 10.8 grams, respectively. If they served no other future purpose—if for some reason they failed to embryonate and to hatch—they had nevertheless given me an actual yardstick instead of a speculative one to use in my clinic. And I *now* felt more than ever skeptical of "E's" loss in weight being attributable to the laying of a solitary egg.

Apparently I had been right in refusing to be misled, for at the third Wednesday session "E" had come up nicely to 439.1 grams—not only a big jump from the week before, but also a gain of 6.7 grams over her initial weight the week before *that*. I briefly considered the possibility that it was not she that was behaving erratically, but rather I who had become flustered in the midst of so much naked chelonian beauty and made a mistaken reading of counters on the triple-beam balancing scale. However, when I remembered with what zeal I conducted all those recordings, I dismissed self-accusations of such frailty. In that case I must accept another vagary as fact also: "D" had not only lost last week's gain but was now below her beginning weight. Thus only "B" and "C" remained as consistent performers, though I must continue to eye them suspiciously, in case it were their turn for being temperamental next.

Perhaps I ought to have concluded that once a week was too often to weigh the turtles. But not being sure what other interval might be better, and having already established the weekly procedure, I continued to follow the original routine. I shall now skip to the morning of Wednesday, June 10, on which I held the sixth clinic, in order to show what sorts of differences had become evi-

dent after that long a lapse. As of that date, "B," "C," "D," and
"E" had gained 26.2, 30.1, 35.0, and 11.4 grams, respectively, over
their bulks at first weighing. If they now contained eggs close to
laying size, all but "E" might be considered to be approaching a
happy event, "B" and "C" each enceinte with three eggs and "D"
with four. "E," for some reason was not in it; perhaps she was too
young, or perhaps undersexed. So I thought sagely, but how
wrong I was!

Late that afternoon I set up my nets for Seaside and Sharp-
tailed sparrows on the salt marsh in Stone Harbor. For best results one prefers a calm day, for otherwise the nets, which are
supposed to be practically invisible, wave in the breeze conspicuously, so that the sparrows manage to avoid them. It had been
calm when I arrived, but soon after I began work in the marshes
a stiff wind rose from the sea and I realized that I might just as
well take the nets down. That is the reason why I happened to
be back at the farm by 7:40 P.M., rather than still disengaging
Seaside Sparrows from tangled meshes in Stone Harbor. And
except for that sea breeze I might remain a sadder but less wise
man to this day.

Automatically I made evening rounds—beehive, new cherry
trees, cocoon boxes, Tree Swallow nests, pigeon loft, turtle pen.
Male "A" seemed to have immersed himself in the pool for the
night—the top of his shell was still accessible to mosquitoes but
he could keep his head comfortably under water except for infre-
quent nips to the surface to take in fresh air. "B," "C," and "D"
had crawled under the pile of dead grass, somehow managing to
pull it over themselves like a blanket. But there was "E" over in
one corner, with an excavation for eggs almost finished. "E," of
all turtles: the one that only this morning had topped the list of
those Least Likely to Succeed!

As quickly as possible I prepared myself for midwifery. I ran
to bring an aluminum garden chair, besides boots, jacket, and a
bottle of repellant—all those being accoutrements against mosquitoes—as well as a notebook in which to record the minute-by-
minute course of labor. Archie Carr's *Handbook* had prepared
me for a long wait. There I read, "Nesting nearly always takes
place in late afternoon, often after sunset, and may extend well

into the night . . . the whole nesting process may consume up to five hours of hard work." Apparently the digging of the hole is the most arduous and protracted part, especially if roots or stones get in the way, but "E" was already near the end of that task, thanks to Rudd's solicitous preparation of the ground when he built the pen. Of course I did not know when she had begun. Even in the most yielding of media, it is no mean trick to dig a hole with your hind feet alone, so this turtle may have been at it for a couple of hours despite our efforts to make things easy.

Thank goodness she did not take umbrage at my nosiness! She had chosen a spot alongside the chicken wire, so that I was able to look through at the closest possible range. At rest, she straddled the hole with her hind legs. To dig, she first had to heave herself clear of the ground until she was resting on a tripod formed by her two front feet and one rear foot; the other rear foot was then free for extension into the hole. The turtle now lowered the rear half of her shell so that the leg in the hole could stretch to the limit of its reach. When the foot had scooped up some soil, it was flexed and everted (Archie Carr uses the word "palming," though with a hind extremity this must be "soling") in order not to spill the load as it was brought to the surface. This was a most extraordinary maneuver, executed with great precision and dexterity. Considering the apparent awkwardness of box turtles in their other actions, I conclude that digging is their one trick that has necessarily become refined in the course of evolution, for individuals performing sloppily in that department were unable to hide their eggs from enemies and consequently left fewer descendants.

The turtle had to raise herself again on the tripod in order to give the emerging foot with its load of soil proper clearance. Now she could collapse from the effort, while the soil was dumped at the side of the hole. The foot and leg then swept it further backward to get it temporarily out of the way, lest it fall into the hole again at the next maneuver. That, after a pause, consisted in a repetition of the former act but on the opposite side, and to do this the turtle had both to execute another push-up and to shift slightly pendulumwise to get her other leg into the hole. Thus she continued with a kind of chain-gang rhythm: heave, shift, leg in, squat, scrape, heave, leg out, flop, push, rest; heave, shift, other leg in, squat, scrape, heave, leg out, flop, push, rest.

I had really been extraordinarily lucky to get there when I did, for the hole was so nearly ready that I was spared the fatigue of watching "E's" travail for more than fifteen minutes. Even by that time, however, I was quite exhausted, for I had found myself grunting with her—or for her—at the beginning of each new stanza of the work chant, and I imagine my face must have assumed a trance-like expression that matched her dreamy one. But when at last her searching feet could not reach any additional soil in their strained groping beneath her—forward, downward, or sideways—she came to a stop and for several minutes simply did nothing at all. Most of the hole's entrance was covered by the rear portion of her body. I would like to have moved her, to inspect the exact shape and depth of the excavation, but naturally I did not dare to interfere. If she had quit now, I can't think what dire things I would have done. As far as I could see beyond the tip of the turtle's tail, the hole extended downward and *forward*, so that the animal was resting partly on its roof. But the opening was rather narrow and seemed to widen deeper down, its neck thus being thick enough to preclude any danger of a cave-in.

Now an egg began to appear. After all the effort of digging, the simple oozing out of this object, with the turtle seeming to be utterly passive, was comical in the sense that comedy entails the unexpected. The egg dropped neatly into the hole, but not so deeply that I could not still see it. The turtle now amazed me by heaving herself up and lowering one hind leg into the hole alongside the egg which she then stroked gently with her claws. She seemed to be feeling it, to find out what had happened. I suppose she had never seen an egg—and still never has, for she did not turn around to look. Archie Carr writes of turtles that arrange their eggs, and even add sand between layers of eggs, but that happens with species that lay larger numbers. Perhaps "E" had that instinct, but simply could not find the right outlet for arranging a single object in an otherwise empty room.

I wrote in my notebook at the time that she "caressed" the egg, though I knew that to be an arrant anthropomorphism. I used the word more to describe the light tactile action of her claws than to define "E's" frame of mind. Anyhow, she with-drew her foot after a couple of minutes and then resumed rest-ing athwart the hole. Five minutes after the first egg, a second appeared, two minutes later a third, and three minutes after that a fourth and final one. She caressed each after it was laid, still without seeming to accomplish any change in their position.

Somehow she knew that she had finished, for she began to cover the eggs at 8:10 P.M.—five minutes after the last one had been deposited and only a half hour after my arrival. Ordinarily I would still have been working in Stone Harbor. Within a short time the eggs could not be seen, and if I had returned to the farm now and made my evening rounds, I would not have suspected "E's" behavior of being more than an effort to scoop herself a small depression for the night. For filling maneuvers were not nearly as stereotyped or elegant as excavating ones; the turtle swept the soil about by describing broad half-circles with her hind legs and soon had it scattered about messily. What fell into the hole did so part of the time only by accident.

The chief item of entertainment being over, I withdrew, for it might be hours before "E" became satisfied that she had prop-erly obliterated traces of her work. There was one thing I must do, however, before allowing myself to go to bed. I must weigh the turtle tonight, to get a true measure of her loss before she had time to take breakfast in the morning and modify the result. On the other hand, I was reluctant to disturb her now, because I felt the welfare of the eggs depended on *her* judgment as to how deeply they should be buried and how firmly the soil should be packed on top of them. Thus I entered a waiting contest with my own patience. Darkness came, and I went out periodically with a flashlight to see whether she had finished, but she remained on duty—doing almost nothing, to be sure, but perhaps doing some-thing *important* at the same time.

At last my patience lost. At 9:45 P.M. the ground over the hole was level enough to please *me*, so I brought "E" in for her second weighing of the day. She had fallen from 444.2 grams in the morning to 403.0—a loss of 41.3 grams *in toto* or 10.3 grams per egg. I wondered whether this difference would still be obvious at the next weighing a week hence—whether I would be able to tell that she had indeed laid eggs, which was the entire primary purpose of the clinic. Thus I did not liberate her now, but kept her in the pen under the same conditions as heretofore.

At the seventh clinic she balanced in at 412.0 grams—a loss of 32.2 grams from the prior week. That was still plenty good enough. I would have erred only in suspecting that she had laid three eggs instead of four. So on June 17 I gave "E" an affectionate pat on the plastron and set her loose alongside the pen, whereupon she began to make a beeline across the lawn toward the Godfreys' farm. I thought that the direction might be just a happen-so, in which case she should veer after a while, perhaps ending up by walking in circles; but she continued straight on. When she reached the highway, I carried her across to save her from being popped by a motorist, and that was the last I saw of her.

My method having proved itself in the case of "E," I felt confident that "D" had had a similar obstetrical experience in my absence during the past week. At the seventh clinic she showed a loss of 28.1 grams from the previous recording. Consequently I let her go at the same time and she, too, headed toward her old headquarters at Godfreys' farm. Kenneth, who happened to show up at this point, said that turtles were a nuisance in their strawberry patch, but I doubt that that could have been the prize my refugees were trying to regain. Strawberries were not yet ripe when these turtles were caught this year. Thus the sluggish-minded beasts would have to remember the Godfreys' patch from last year, and also sense that strawberries were now in season. I did not credit them with reactions that acute.

My indolent male now seemed rather superfluous, for the remaining two females must by this time contain shelled eggs, and if he had not already fertilized them it had become—alas!— too late to rectify that matter. Consequently I released him, too. He rather confused the pretty picture of homing turtles by making straight for the woods, *away* from the highway and his old home beat. He simply had never been very bright.

Since I knew the exact spot where "E's" eggs were buried, I could see no point in digging up the rest of the area of the pen in a search for "D's" nest. If I decided eventually to watch an egg or two in the house, I could dig at "E's" site and leave the rest of the pen undisturbed. Presumably "B" and "C" would eventually lay, and after they had been released I would rim the base of the chicken wire with a strip of solid metal sheeting six inches high or so to prevent babies from escaping through the meshes. Then it would be interesting to see just how long I could refrain from excavating the whole show.

But meanwhile I had my caesarian eggs to watch, and it seemed as if something propitious must be happening. The fact that they were behaving differently made it look as if one of them might be alive and the other dead. Perhaps it is inaccurate or unacceptable to speak of the behavior of an egg, but at least let me say that they changed from day to day. When Kenneth first brought them, their shells were not white but looked more as if their pores were still filled with maternal moisture. One egg remained in this condition, with an appearance of being covered with skin or membrane rather than shell. It definitely looked unhealthy, being of a dull yellowish flesh color. The other gradually cleared up and began to look like other kinds of turtle eggs that I have seen. However, it—in a most odd fashion—*began to gain weight*. An egg can't eat, and the only thing I can propose is that it must have been absorbing fluid from the humid atmosphere in the jar. To show that this was not my imagination, I herewith present the weights of the two eggs as obtained under identical circumstances on various dates:

	"Skinned" egg		*"Shelled" egg*	
May 24	10.8	grams	10.5	grams
June 5	10.8	"	10.5	"
June 13	10.8	"	10.8	"
June 25	10.7	"	11.2	"

The question was: Which egg was alive? I was betting on the shelled egg. However, I had already lost one box turtle bet ("E's" pregnancy), so I was not yet shouting my opinion. The eggs were now a month old. If either *were* alive, I should soon be able to see some signs of an embryo or blood vessels by shining a strong light through it.

I was fortunate in having two mothers, "D" and "E," whose weights indicated their lying-in so clearly, because "B" and "C" failed to show contrasts of the same degree. By the time of the eighth clinic, on June 24, I felt fairly sure that they, too, had laid, but their detectable losses were each less than 20 grams. They may have gorged themselves promptly after nesting and thereby obfuscated their clinical histories. I decided to keep them awhile longer, just to be sure. But whatever they might already not have done—or might not do in the future—I was more than content with the knowledge that I possessed two undoubted box turtle clutches.

It is always a good idea to go back and browse over what you have read before. That exercise modifies many an incorrect impression—at least in my own case. Now I found in Archie Carr's book that while a great deal of vague information exists regarding the incubation periods of various kinds of turtles' eggs—including those of the box turtle—an occasional terse and very positive statement can be located also. After calling the box turtle's period extremely variable and giving its limits as 70 to 114 days, Carr goes on to quote the belief of H. A. Allard, who published scientific articles on this species over a span of forty years, that the average is between 87 and 89 days. The latter now struck me as a definite measure that I could use for schooling my expectancy. Eighty-eight days from June 10, when "E" performed her blessed feat, would be September 6, and I must somehow gear myself to patience and restraint while subterranean embryos worked toward that date of release.

Those were good intentions. On the other hand, I must confess what really happened. On July 15 I found that "B" had fallen *below* her initial spring weight, while "C" had declined almost to that level, so I let them both go. The turtle pen was now empty and desolate, except for its verdant growth of morning glory and dwarf ragweed. Come to think of it, what would the roots of those plants do when they reached turtle eggs? Obviously they would not turn into parasites and penetrate them, but could they perhaps form constricting masses that deformed and strangled such yielding objects? Turtles choose sandy spots in preference to overgrown ones for laying. Is that for more than the obvious reason that the digging is easier?

I can't pretend that this worry was the sole excuse I had for digging up the eggs. Possibly it would not be fair to give it more than 5 percent of the credit. That covers me with a shameful 95 percent of irresponsibility. And now that I've admitted it, I still don't feel any better. On July 19 I picked up a short-handled shovel and got into the pen on my knees, with the intention of digging carefully spadeful by spadeful until I came finally to the site of "E's" known cache. I began rather carelessly, somehow convinced that you never strike gold at first, and in an absent-minded way that I can't explain now, thinking more of future baby turtles than of turtle eggs directly underground and close to my hands.

Then it happened—almost immediately. My shovel turned up a nest of eggs that I saw with dazed horror were neatly sliced in half. There were four of them—just as in "E's" case—and they were fertile: I could see the tiny embryos with their bulging eyes. For a moment I did not grasp the enormity and finality of my negligence; I thought that somehow I could turn back those fifteen seconds and live them again with better considered caution. Then as I watched the broken yolk sacs yield their contents to the sandy earth, I experienced the shock that always comes with reminders of the reality of death, that inevitable state which we so fervently try to believe is only imaginary.

But wasn't this overreacting? I still had "E's" nest to fall back on, and there was also the rest of the turtle pen to dig up, just in case "B" and "C" had dropped an egg or two after all. I stopped woolgathering and resumed digging, with an appropriately light touch this time, albeit with a foreboding that somehow the whole project had now been botched. Nor was it a bit comforting to find that moles had tunneled everywhere inside the pen. The chicken wire had been countersunk deeply enough to prevent turtles' burrowing out, but not to exclude true miners. As yet I do not *know* what a mole would do if it met a buried mass of eggs in its travels, but my guess is that those particular eggs would become a snack.

At last I reached "E's" final square foot of earth, without having found any other eggs. Now I carefully lifted up the whole block so as to avoid cutting anywhere into it. Then I was able to crumble the mass in my hands, working carefully toward the exact spot where "E" had labored. At first it seemed as if nothing were there, and I wondered fleetingly whether the whole episode today might gratefully be only a passing flight of madness. But then I found them, four collapsed eggs overgrown with mold. Perhaps they had been sucked by moles, but in their presently long-extinct state it was not possible to be sure how they had been violated.

So that was *it*. I clambered out of the pen and regarded its completely excavated interior in a sort of doubting reverie. How could my plans have gone *totally* wrong? I am used to less than perfection in my pursuits. Indeed, I am sometimes happy to reap as little as half, one-third, or even less from what I have set out to harvest, whether this be material gain, such as reared cocoons from a fertilized Cecropia moth, or simply data in my notebook concerning nesting successes of Seaside Sparrows. But usually there is at least *something* to show, no matter how trivial the result, and I now felt that I must have overlooked some remnant that would rescue me from utter failure. Was there not one baby turtle somewhere, one egg—?

Ovum ex machina! I still had one of Kenneth Godfrey's Caesarian eggs in the humidified pickle jar in my dining room. If I had accepted those eggs condescendingly on May 24, I must now revise my attitude to one of reverent gratitude. The second egg—the one losing weight—had sprung a leak after several weeks; it must have suffered an invisible crack when the power mower delivered it, and subsequent decay could be attributed to bacterial invasion. Anyhow, it went all to pieces. But the "gaining" egg had me puzzled. At times I was quite happy about it and felt that it was doing what a well-behaved egg ought to do. But then it went on to peculiar things that to me were unheard-of and heretical in properly regulated egg societies.

What I considered good was the appearance of blood vessels that I was able to see by shining a strong light through the egg on July 1 when it was thirty-seven days old. This corresponded in general to what my successful mud turtle had done three years ago. But *that* egg then proceeded to entertain me further with the antics of its embryo, with the development of an air space at one end, and with a regular and gradual *loss* of weight that accorded with the notion that it was using up or metabolizing part of its substance during the developmental process.

Not so Kenneth's box turtle egg. After the first of July it became opaque, so that I could no longer see blood vessels, and I never did detect a moving embryo or an air space. Perhaps it was dead and bacteria had rendered the contents turbid. This seemed to me to be no idle fear, for I could imagine that only something dead might become bloated by soaking up moisture in a reverse direction from the natural biological osmotic flow. Growing? Indeed the egg not only gained in weight, but its linear dimensions increased also. Fortunately I had continued to weigh the eggs each Wednesday when I held box turtle prenatal clinics, so I had some nicely spaced data that traced the unexpected changes. Weighing both eggs was fortunate, too, because the widening difference between them proved that I was not simply making a series of mistakes with either one. I did not re-measure the hopefully viable egg until I had become convinced of its increased bulk. Then I realized that the law of conservation of mass must perforce mean enlargement of dimensions, and by trying to look at the egg with a fresh eye, I persuaded myself that it was no longer as cylindrical and reptile-like as it had been but now resembled an avian ovoid. But here are my complete data, and to hell with a fresh eye that sees what it wants to see:

	Egg A	Egg B	
Date	Weight (in grams)	Weight (in grams)	Dimensions (in millimeters)
May 24	10.8	10.5	33 x 21
June 5	10.8	10.5	—
June 13	10.8	10.8	—
June 25	10.7	11.2	—
July 1	10.4	11.5	33 x 23
July 9	Discarded	11.7	33 x 24
July 15		11.8	33 x 24
July 22		11.6	—
July 29		11.5	33 x 24
Aug. 5		11.3	—

The increase at its maximum on July 15 was 1.3 grams or 12.4 percent in excess of the original weight. By this time I wondered if the egg was about to burst. But I wondered some other wonders as well. Could it be that the egg was still imperfect when the power mower exposed it? Would it normally have absorbed further moisture from the walls of the maternal reproductive tract until it reached its present bulk? I would not have thought so, for it is my understanding that the shell is not applied—at least in birds' eggs—until the central yolk and outer albuminous layers have been fully laid down. On the other hand, do naturally born box turtle eggs gain weight too? Is the capacity to take up water an acquired safeguard to help buried eggs survive droughts? If only I had robbed "E" of, say, two of her eggs that night and fostered them in the pickle jar along with Kenneth's!

But after July 15 a reverse trend set in. The egg began to lose. However, at the same time it remained dark. Could a dead egg move both ways, first in one direction and then the other? Allard's eighty-eight-day incubation formula, by the way, gave August 20 as a calculated end point, though since the power mower had intervened ahead of the normal laying date, I could easily conceive of a considerably delayed hatching time, perhaps not until September.

"Julius Caesar" was born on August 12, 1970, after an eighty-day incubation period. He had large dark eyes. His head was dark, with a white beak and a sharp little white egg tooth. On the next day, still sporting an umbilical stalk, he weighed 8.7 grams.

Chapter 7

THE MIGHTY SPARROW

I have often thought of asking my family and friends to knock me in the head if they see me taking up another hobby. But I always keep forgetting to make the request. Anyhow, so far as sparrows are concerned, I doubt that the responsible people would have known the time to wield baseball bats or lead pipes, for I hardly realized myself when it was that those puny birds had secured me inescapably by their mighty attraction.

It all began on a little sandy side road off the causeway, near Stone Harbor, where I was setting large-mesh Japanese mist nets to catch various kinds of sandpipers and plovers. I had never banded many birds in the category of waders, but such shorebirds often frequented a few shallow marshy pools near the road. The spot was a handy one at which to work because I could sit in the car while watching the nets close at hand. Birds seem unable to recognize human shapes within the larger outlines of automobiles. Since a car in itself does not signify danger to them, I could use the sedan as effectively as if it had been an ornithologist's conventional hide or blind.

The trouble was, I caught very few sandpipers and plovers. My experience with bird nets showed me at once what was wrong, but there was no way to correct matters. When dealing with land birds, I always tried to place the nets against a relatively dark background, such as the edge of a woods, a fencerow, or shrubbery of any sort. A net then became inconspicuous, and unsuspecting birds readily flew into it. But out here in a salt marsh there was no way to lessen the impact of a net on the eye; in fact, it stood out as a conspicuous alien object, despite the thin black strands of its mesh. Moreover, the open face of the marsh gave the slightest breezes unconfined license, so that even on what seemed to be calm days the net would be constantly waving back and forth like a banner, further proclaiming its presence.

Oh, I did catch a few shorebirds, though not enough to justify my trouble. Sometimes one would be chasing another. Presumably the one in front was looking back over its shoulder, while the one in pursuit had eyes only for its target, and they therefore both piled into the net with one almost atop the other. On other occasions I would slyly drive feeding birds ahead of me on foot, gradually getting closer to the net, when with a sudden rush I would startle them into heedless flight directly into the trap. But usually they became alarmed before we had gotten near enough for that ruse.

It was on one of those chases that I stumbled on my first Seaside Sparrow's nest. Fortunately "stumble" is only a general term in this instance. My foot was still within a final step of making contact with the nest, when a sparrow buzzed into the air just ahead of me. It flew forward in a straight line at no more than my shoulder height and then dropped back into the marsh grass about fifty feet away. I had been enjoying Seaside Sparrows, for as I sat in the car there would often be a singing male nearby on a weed stalk or even a grass tussock only an inch higher than neighboring tufts. I was aware also that they must be nesting here; those males were announcing territories as plainly as if they spoke English. But whereas all creatures in nature are interesting, I simply haven't enough time to be *intensely* absorbed in them all, and there had been no previous trigger to set me off on a sparrowy trajectory.

I *had* thought vaguely about looking into the doings of those birds. However, I just as vaguely put such ideas aside because of having read that Seaside Sparrows' nests are exceedingly hard to find. Besides, it was already after mid-June when I began my sandpiper studies, and I assumed that nests would be empty by this time, though with some fledglings perhaps still being fed by their parents on a mobile basis in other recesses of the marsh grass.

But here was a nest containing four eggs on June 30th. And as for finding it, I still could not say whether the feat had been hard or easy. From my point of view, I could state that I had done nothing, for I hadn't been looking for nests. The work had all been chalked up by the bird. On the other hand, if I could duplicate the performance, with volition coming solely from me, I might have some right to credit for the find. So thinking, I first marked the nest by driving a small driftwood stake into the mud a few feet away, and then found a second nest after searching through the grass for all of five additional minutes. *That* must be more than beginner's luck.

These two nests were typical of many others I have seen since that day. In fact they might be called typical of nests considered as a genus. What comes to your imagination when you think of a bird's nest but a deep cup built of swirled grass stems? What kind of eggs does it contain if not heavily speckled ones? The chief distinction these nests had was their position. It *would* have been very easy to tread on them, had not the first sparrow delayed its exit until I was almost upon it and put me then and henceforth on the alert with every step I took in the marsh. The grass can hardly be thought of as having strata; yet the nests neatly occupied a middle storey between earth and sky, with lower parts of the stalks providing foundation to keep the structures raised above the mud, and terminal blades arching overhead to blur the contents from prying eyes of gulls and other flying scavenger-predators.

The discovery that incubation was still going on immediately made the thought of netting and banding the parent birds a sensible one. Banded birds as small as sparrows are only very seldom found dead by the public. Therefore a bander must resign himself to a heavy work load for small return if he catches birds only randomly on their travels. He may have to report using a thousand or more Fish and Wildlife Service bands before that agency can notify him of a single recovery. But if he works on a nesting ground or at a fixed feeding station or any other permanent location possessing some sort of attraction for birds, he automatically sets up his own system for being both the bander and the subsequent recoverer. A Seaside Sparrow nesting in a salt marsh near Stone Harbor in 1967 might possibly nest there again in 1968—and in 1969 and 1970, and so on.

It suddenly seemed likely to me that scientists had not yet bothered to make any such study of Seaside Sparrows. Why should they, when there were still plenty of more attractive birds inviting graduate students of ornithology to work out their life histories? Not only were those other birds prettier, but they lived in pleasant woods and fields instead of muddy marshes where tides stranded dead sharks and fiddler crabs clattered over the slime. Seaside Sparrows were appropriately of a grayish mud color. They had a small yellowish spot before each eye and a blackish moustachial streak bordering a whitish throat. Their short tail feathers were sharply pointed, giving the tail a ragged, bristly look. Short wings, big feet, a long unsparrow-like bill: that finished the portrait of this unpromising creature.

Of course to me they were beautiful. I find even a castaway shark's carcass attractive in the proper setting. Consequently I was not dumfounded when what had at first been merely an indefinite desire to know more about Seaside Sparrows began to feel more like a compulsion. Then in a cascade I saw a torrent of reasons why these particular sparrows should command my special interest. Wherever I had studied in the South Jersey marshes, they had sat up and repeated their choppy ditties. By a tolerant stretch of kindness one could call the buzzed notes a song. But the sound was constantly there as background music or noise—on Cedar Island where Ospreys shrieked and Clapper Rails coughed; in whatever marshes hordes of *Aedes sollicitans* larvae drank the soup of their incubators; alongside diamondback turtles when they hauled out of tidal channels to lay eggs. Moreover, I could think of a whole index of investigations to make. Banding should reveal more than the steadfastness with which some birds might return to this particular marsh year after year. If I could follow the sparrows' individual careers, I would learn the secrets of their homelife: whether the same males and females remained mated throughout one season and in successive summers; whether one or both parents took part in incubating eggs and feeding nestlings; what size the nesting territories might be, and therefore how many families this little marsh could accommodate; and so on through the standard format of a conventional life-history thesis.

But wait. First let's see whether I can succeed in putting salt on Seaside Sparrows' tails. If this species were endowed with special faculties that aid it in perceiving and avoiding Japanese mist nets, I would not be in possession of a project after all. Perhaps that would explain why other ornithologists had neglected the birds: it was *not* their plainness or the marshy effluvia. Unfortunately my sandpiper nets had meshes too large to retain sparrows, so I could not make an immediate test on the spot. After marking the site of the second nest with an inconspicuous weathered builder's lath driven into the mud several paces distant, I hastily folded the nets and gathered up my poles, in order to race home and prepare for the switch.

Of course life is not as simple as that. I mean, there is more to it than just sparrows and me. Frequently I have responsibilities to other hobbies, and occasionally even people infringe. During the following two weeks the best I could do was to visit the marsh briefly a couple of times to check on the nests, but there was not enough time to set up bird nets.

At length, on July 14, I set up three nets and caught four birds—three Seasides and one Sharp-tailed Sparrow. In a way I am entitled to say, "I knew it all the time," because Sharp-tails, though more widely distributed than Seasides, occupy a practically identical habitat where the ranges of the two species coincide. Sharpies are generally found in somewhat drier spots, but as the birds come and go in the environment, they both end up covering the whole area. Sharp-tails are smaller, browner, and more heavily streaked than Seasides, but their secretive habits among grass stems and their short buzzing flights when briefly chased into the air are the same. Thus unless I had taken special trouble to study each sparrow that had flown up while I was driving sandpipers, I could not have distinguished between the two. But having heard and seen the perched, singing Seaside males, I carelessly called them all Seasides.

Well, that made no difference. In fact, it was all the better, for if Seaside Sparrows needed to be studied, the same was probably true of Sharp-tails. And if the two could be investigated simultaneously with a single set of equipment, it would be wasteful not to avail myself of the proffered extra gift. Nevertheless the situation was now a bit more complex than it had been, and as I banded the birds I began to foresee some special provisions that ought to make for next summer's successful nesting program.

Some of the items of interest I have mentioned, such as determining which parents incubate eggs and at what times they do it, as well as learning similar information about the feeding of nestlings, would entail catching the birds so often to read their band numbers that they might desert their nests. If not that, the interruptions might be deleterious to the eggs and young. Besides, my constant attention to certain nests could reveal their presence to a predator which forthwith might wait for my absence in order to make systematic observations in its own behalf. Goodness knows I enjoyed handling the birds—to me that is one of the attractions of birdbanding—but in this case my objective should be to find out as much as I could with the least possible amount of physical contact.

It was my good luck to have been invited only a few months previously by Professor David E. Davis to visit his home to see his scheme for studying birds in relation to mosquito-borne viruses at the Pennsylvania State University. On one morning during my visit, two of Dave's graduate students in ornithology had taken me to a marshy meadow where I was shown their methods of learning the very things about nesting Red-winged Blackbirds that I now wanted to determine for sparrows. They told me that they had overcome the excess-handling problem by color-banding the birds. This technique was not new, but they had added a novel facet to it. Previous workers had used celluloid bands of various colors in addition to the regulation aluminum Fish and Wildlife Service band in order to recognize individual birds at a distance by the color formulae they wore —yellow, red, or blue on the right or left leg; and even the position of multiple bands—yellow above or below red, etc. But recently the supply of colored bands had run out in the United States and it was now necessary—and costly—to get them from England. These young men had found that they could much more cheaply wrap a narrow strip of colored plastic tape in various positions on a blackbird's right or left tarsus and achieve perfectly satisfactory identification marks. We looked at a male Redwing at the top of a low tree through binoculars and could easily see a bit of yellow tape that they said had been applied a year ago.

Just as it is necessary to get a license to band birds from the Fish and Wildlife Service, United States Department of the Interior, so that august authority must be approached also for permission to undertake color-banding studies. In the first case the agency has to be convinced that the applicant can identify birds accurately, but in the second it wants to know why the bander is interested in the more meticulous recognition of individuals. If it is only so that he can amuse himself by telling Cardinals at his birdfeeder apart, the application will not get very far. I have, incidentally, often thought how fortunate I have been to live during this century, for surely a day is coming when *all* applications, including those for ordinary banding, will be turned down because every bird, including the rarest species, has been studied in full. I did not know quite how to phrase my letter, since I was not a graduate student, I did not hold a faculty or a curatorial position, and my interest was really somewhat akin to a curiosity about Cardinals at birdfeeders. Somehow I succeeded in convincing somebody, for early that fall, long after the 1967 nesting season was over, I received official approval to color-band Seaside Sparrows next year.

That winter I spent in Trinidad at one of my former places of employment, the Trinidad Regional Virus Laboratory. After a bit of wangling I managed to be accepted as a paying lodger and boarder at Simla, the tropical field station of the New York Zoological Society in Trinidad's fabulous Arima Valley. And here, to my joy, I resumed an association with a young ornithologist, Dr. Charles T. Collins, who, when I first met him at Simla several years previously, had been a graduate student just such as I devoutly hope all ornithological graduate students are. As soon as I could get in a word, I began to brag about my sparrow project —and crash! he disclosed that it had already been done by another young ornithologist, Dr. Glen E. Woolfenden. It happened that I had met Glen at an ornithological convention only a few years ago; he did not mention sparrows then, but I suppose there was no reason for him to have. They certainly were not haunting my brain at that time. However, Charlie now told me that Glen had chosen Seaside and Sharp-tailed Sparrows as subjects for graduate study and had worked with them actually *in New Jersey* somewhere, almost in my backyard. Then he had written and published an account of his findings, exactly in the manner that I contemplated.

What should I do now? I did not want to turn in the precious color-banding permit. Yet I felt reluctant to plod step by step through a project that had already been covered by someone as able as Glen Woolfenden. Obviously I ought first to read Glen's paper. Perhaps there were still some problems that he had not worked out. Maybe he had purposely omitted certain phases that I could now fill in.

On my return to the farm in March, I ascertained that Glen was currently on the staff of the Zoology Department at the University of South Florida in Tampa. In his reply to my query he kindly commended my proposed study and then directed me to write to the Museum of Natural History at the University of Kansas in Lawrence for a copy of his work, since it had been published at that center. This had been in 1956, longer ago than I would have suspected, and it was possible that the museum had run out of reprints. That would mean that I must somehow get a copy on microfilm, and in such case maybe I really would desist and leave the beleaguered sparrows to themselves.

But some weeks later I received the pamphlet: "Comparative Breeding Behavior of Ammospiza caudacuta and A. maritima," by Glen E. Woolfenden. It was about twenty-five pages long, and included charts, tables, drawings, and photographs besides the text. The contents listed the following divisions: Introduction; Materials and Methods; Description of the Area; Flora; Reptiles; Mammals; Predators; Passerine Associates; Winter Status and Spring Migration; Territory; Voice, Copulation; Nests; Eggs and Incubation; Young; Food, Feeding and Care of the Young; Acknowledgements; Summary; Literature Cited. Exactly as I would have done it, if only I had thought of it first!

By the time I had read the treatise carefully two or three times, I could begin to see a few variations between Glen's study area and mine. He had worked farther up the coast, near Lavallette in Ocean County, and though the marshes there were coastal, like Stone Harbor's, they did not seem to be quite as salty as our local ones. His supported cattails, and the cattails attracted Long-billed Marsh Wrens, while that plant and those wrens were both lacking in Stone Harbor. Moreover, his marshes were more heavily diked, with an abundant growth of marsh-elder bushes along those ridges. Seaside Sparrows nested under the marsh-elders, whereas my evidence thus far was that at Stone Harbor they nested out in the open expanse of marsh grass.

However, those may have been minor or entirely trivial points. Such differences could scarcely affect the general pattern of the birds' life histories. For example, Glen discovered that although male Seaside Sparrows take no share of incubation duties, they do help feed their nestlings. On the other hand, male Sharp-tailed Sparrows take no part in domesticity at all after mating. Those observations, true at Lavallette, must surely be duplicated at Stone Harbor. But I was finally left with one challenge: Glen had failed to determine the incubation period of either species. At last, here was lore to be learned. No matter how trifling such grains of information might really be, I at once magnified them into mountains. Henceforth my quest would be to conquer them.

There remained a tiny annoying possibility that Glen Woolfenden's work had been amplified or superseded more recently by some ultramodern expert, unknown even to Charlie Collins. The last word in such matters could soon be found, when the final three volumes of Bent's *Life Histories of North American Birds* appeared. This classical series, begun in 1919 by Arthur Cleveland Bent of Taunton, Massachusetts, and published by the Smithsonian Institution of the United States National Museum, had extended through twenty volumes by the time of the author's death in 1954 at the age of eighty-nine. Those tomes had covered all North American bird families north of Mexico except Fringillidae, that great group encompassing cardinals, grosbeaks, buntings, towhees, finches, and sparrows. The work of finishing this part of the series had been undertaken by Dr. Oliver L. Austin,

Jr., at the Florida State Museum, University of Florida. Following Mr. Bent's practice, Dr. Austin served variously as author, editor, and compiler in bringing together the knowledge about fringillid life histories, in some cases piecing together fragments of information contributed by many people from scattered places over many decades, in other instances inviting a noted authority on a particular species to submit a complete life history report under his own name as the author.

Undoubtedly there were many ornithologists besides me who were impatient to receive volumes 21, 22, and 23 of "Bent," as it was still popularly called. But I am certain that no one turned as promptly to page 819 as I did when I unwrapped those 1968 treasures. Who was better qualified to have written about Seaside Sparrows than Glen? No one, of course, so he had been selected to do it. Apparently there was another Sharp-tail pundit around, for that bird's life history had been authored by Dr. Norman P. Hill of Fall River, Massachusetts. And on page 802 he stated, "My data from Barnstable, Mass., indicate an incubation period of 11 days." Thus one mountain had been removed already. However, Glen reported that the Seaside Sparrow's incubation duration was still unknown.

There was no telling how many eager people might now be trying to fill that gap. Obviously it *would* be filled soon, just as the Sharp-tail had yielded its secret since Glen's negative study in 1956. Then why not sit back and wait until the news appeared in one of the technical bird journals? Was I really only curious to know a fact of nature, or was I competing for the distinction accorded a discoverer? Dispassionate science is not cognizant of individuals, but please show me a dispassionate scientist. Do you suppose that Columbus would have been satisfied if Queen Isabella had thanked him for his hunch that the world was round and then commissioned someone else to verify it?

It is fruitless to argue about my motivation. More to the point, I felt that I had now established a sufficient background to set about my work with a proper sense of direction. The very first step, logically enough, was to look into the matter of colored plastic tape. Since I did not foresee the need—or the possibility, either—of banding large numbers of sparrows, I felt that two

colors would be plenty. It was easy to begin a chart that showed
how high the combination of two colors and two legs would go.
Thus:

Bird No.	Right Leg	Left Leg
1	Yellow	–
2	–	Yellow
3	Yellow	Yellow
4	Red	–
5	–	Red
6	Red	Red
7	Yellow	Red
8	Red	Yellow
9	Yellow over Red	–
10	Red over Yellow	–
11	–	Yellow over Red
12	–	Red over Yellow
13	Yellow over Red	Red
14	Yellow over Red	Yellow

and so on to the point of observing about two dozen sparrows of
each species, for I felt I might as well study both kinds, despite
Dr. Hill's scoop, and I could therefore use each color combination
twice.

I looked for colored plastic tape in several drug stores and
five-and-tens without seeing anything suitable. Unfortunately I
had neglected to ask Dave Davis's graduate students what brand
they used, and since they had not mentioned a particular make as
being important, I had assumed that there would be no scarcity
of appropriate kinds. Finally I spotted a promising-looking product in a commercial stationers. While it was too wide, I could
easily cut strips to size with a razor blade, and it did come in
bright red and yellow shades that should be conspicuous against
a background of marsh grass.

Dave's students had worked in a freshwater environment
and their tapes had stayed on for a year. I wondered whether
sea water, or at least a highly brackish medium such as often inundated the Stone Harbor breeding area, might not cause the
gummy substance of the tape to deteriorate more rapidly. I was

counting chiefly on the tape's sticking to itself, that is, I intended to wrap it around a bird's tarsus at least twice. Thus the most important adhesive surface would be sealed in and the plastic band ought in theory to stand up for a long time. Nevertheless it would be wise to make some preliminary tests of that assumption. I did the easiest one first, using ordinary tap water at the kitchen sink. Having wrapped a couple of matchstick-sized twigs with both red and yellow tape, I left them to soak overnight in a saucepan. I did not need to wait that long. Within an hour they had uncoiled themselves and were swimming around as flat ribbons.

Meanwhile I had realized that I must check another angle, namely, the distance at which I could read color combinations through binoculars. My bird-glasses were of the conventional 7 x 50 variety, excellent for everyday bird-watching. But those little twigs (before I gave them the water treatment) revealed their red and yellow markings only at the closest focal range of the instrument. Even if I could manage occasionally to get that near to a bird, I should indubitably be disturbing it and thereby would negate my entire philosophy of noninterference. And that is how I came to spend two hundred dollars. For years I had wanted a 50-power "bird-spotter" telescope, such as big-time birders use, but I could never have bought one with an easy conscience.

By the time I finally located the right kind of plastic tape, and had waited the necessary number of weeks for the telescope to arrive, the nesting season of 1968 was well advanced. I had lost the important weeks during which most nests were provisioned with eggs and I must consequently defer my study for another year. However, I could at least try out the tapes and the telescope. Fiasco! The adhesives did a splendid job clinging to sparrows' legs and to themselves, but even with a telescope I could barely see them. It was not because of any fault in the scope. Rather, the sparrows simply never sat where I could observe their legs. Even singing males, perched motionless on tufts of grass while they repeated their chants for ten or fifteen minutes at a time, usually exposed only the upper half of their bodies, so that under those most favorable conditions I still could not glimpse the red or yellow flecks.

Well, I had the telescope and that was something gained. I have since used it happily at the Brigantine Wildlife Refuge where it picks out distant ducks admirably. Likewise I have seen the rings of Saturn through it. As a matter of fact, Glen's paper had taught me a lesson that made the whole color-banding scheme rather unimportant after all, so I was not really daunted by the sparrows' adverse secretiveness. One of the chief uses of color was to have been sex determination, since male and female Seaside Sparrows are identical in appearance. If I could have read the color codes of singing birds, that would have revealed the male population, and I could make corresponding notations opposite their Fish and Wildlife Service band numbers in my records. The nonsinging half of the population would presumably have been comprised of females. But Glen drew my attention to an entirely different method.

Just prior to the period when he began his studies in Lavallette, a remarkable article (which I completely overlooked at that time) appeared in *The Auk,* official journal of the American Ornithologists' Union. In this communication Dr. W. Ray Salt, of the Department of Anatomy, University of Alberta, Edmonton, Alberta, pointed out that a protuberance appears at the cloaca, or vent, of the males of many species of passerine birds during the breeding season. This bulbous swelling had already been mentioned in print from time to time by other ornithologists, but they had not recognized its significance—indeed some of them had considered it a pathological abnormality. With his anatomical background, Dr. Salt sensed that the protuberance might have a true function. On making a number of dissections, he found that the body was not pathological at all but owed its existence to the greatly distended lower ends of the two genital ducts, each serving *pro tem* as a seminal reservoir. After the breeding season was over, the ducts regressed to normal size and the protuberance disappeared.

Now it was a race for other investigators to see what additional birds they could find with the new "organ." Dr. Salt had made a special study of it in the Vesper Sparrow, but besides that species he said he had observed it in several other fringillids including the Sharp-tailed Sparrow, as well as in Yellow-throats,

Western Meadowlarks, Yellow-headed Blackbirds, Redwings, and Cowbirds. As might well have been expected from the scope of that motley list, Glen Woolfenden quickly signed up the Seaside Sparrow for its share of cloacal notoriety.

Actually there had been another available physical trait waiting for me to use in distinguishing between the sexes of Seaside Sparrows and that was the female's brood patch. This characteristic is not always as distinct as one might wish. By itself it would be a useful though sometimes an inconclusive guide. The brood patch is an area of skin on the abdomen of incubating birds where a temporary increase in the flow of blood enhances the transfer of heat from the parent to its eggs. The patch shows up as an unusual extent of nakedness, the bird appearing to be potbellied. When both sexes of a species incubate, each parent develops a brood patch. Still, this is chiefly a relative change and there could easily be some intermediate cases in which one would hesitate to diagnose a bird positively as being engaged in nesting. Thus I might find the trait of only limited value in identifying female Seaside Sparrows.

But in the absence of a cloacal protuberance, even minimal brood patches suddenly became significant. Indeed it was remarkable how Dr. Salt's article now changed my entire concept of look-alike birds by endowing the two sexes with distinctive external genitalia, almost like mammals. Of course this was true only of birds in the hand and during the breeding season, but it nonetheless brought them into line with showy species in which males could be distinguished at sight from females by their bright colors.

Knowing the sex of my birds would help in more than mere studies of incubation. By setting up nets in various parts of the marsh at different times, I would now be able to tell how far the members of a pair roamed during periods when they were occupied with home duties. After the young had been fledged, I could detect if and to what degree such patterns changed. I might learn whether mates left the marsh at the same or different times when summer was over, and what was the sequence of their arrival in spring. Furthermore, if some unlikely sparrow were found dead in a Georgia or Florida tidal marsh during winter, the Fish and Wildlife Service would have gleaned an extra iota of information from the record if its computors could associate that event definitely with one or the other sex.

I gradually realized that a year-to-year study of banded Seaside Sparrows in this marsh would be a novelty in itself. Glen Woolfenden had been occupied in Lavallette for only one breeding season, and apparently none of the other observers he mentioned in "Bent" had ever banded any birds. Thus it was not yet known whether any of the same sparrows return to a given marsh in successive years. While I did not need to follow large numbers of birds to gain information about their nesting behavior, the solution to questions regarding long-term tenancy of a breeding area entailed statistical measurement of whole populations, and that did require relatively large figures to achieve significance. Hence I must give some kind of boundaries to my study area, then map it, and finally make a conscientious effort to band every sparrow living within its confines.

My 1967 tally of banded birds had totaled only 8 Seasides and 1 Sharp-tail. In 1968, when I added color-banding to the scheme, my score was only slightly better: 15 Seasides and 6 Sharp-tails. And though I had set my nets at the same general locations in both years, I had not recaptured any of the 1967 birds in 1968. However, a statistician would promptly pat me on the back and tell me not to be discouraged, assuring me that 9 birds in 1967 was far too small a sample for me to place confident pessimism in, so far as their not returning to Stone Harbor was concerned.

As matters stood at the beginning of 1969, then, I had two projects to pursue: One specific—concerning the exact length of incubation—and the other more general, dealing with a special phase of the migratory movements of populations. One might think that the question of incubation could be settled very quickly. In fact, one could wonder why it had not already been disposed of. Indeed, why had not Glen, as the foremost Seaside Sparrow expert, done so? His excuse was that when he arrived in Lavallette to undertake his work, the nests were already full, and obviously you have to know when the eggs were laid to know how long it takes them to hatch. But there are other difficulties.

Oh my, yes! There are *lots* of other difficulties.

Offhand it would seem to be easiest to collect an egg the instant after it was laid and then hatch it out in an incubator, measuring to the hour, if not to the very minute, how long the developmental process had taken. True, that would give a minimum incubation period (and I am assuming that the incubator was set at the same temperature that the bird would have provided for its eggs). But what the ornithologist is after is knowledge of the time it actually takes to hatch out an egg in nature. Would this answer be the same as one furnished by the incubator? Among some species it might be extremely close. Male and female pigeons and doves, for example, spell one another on the nest, so that the eggs never become chilled from the time incubation begins. But birds such as Seaside Sparrows, in which the female leaves the nest unattended whenever she must feed herself—which is several times a day—subject their eggs to unavoidable exposure during those absences. Since the environment is usually colder than a bird's brood patch, heat is lost while the parent eats.

A bird's egg is poikilothermic—cold-blooded—like an adult reptile or amphibian, having no means of regulating its rate of metabolism to control the generation of heat. Nor has it any mechanism, such as insulating feathers and fur, to prevent heat dissipation. It accepts heat from the brooding parent quite passively, but its response is nevertheless in strict accord with the energy thus absorbed. At parental blood heat, internal development of the egg proceeds at full speed, but at each degree lower than that maximum, vital processes are proportionately slowed. Thus even a five minute's absence of the female Seaside Sparrow on a coolish day imposes a delay on hatching time. Moreover, embryos are able to survive lengthy periods of neglect, especially in their earlier stages.

Where does that consideration lead? For one thing, it is inevitable that no two female Seaside Sparrows take off exactly the same number of minutes during the total period of incubation. Nor would environmental temperatures be exactly the same at any two unattended nests. Therefore the incubation period that I was seeking to learn must be a variable quantity. How widely could it fluctuate—by an hour? six hours? a whole day? two days? Answers to these questions would accrue only after I had obtained information from several nests.

But here another problem immediately arose. In order to know when the eggs were laid and when they hatched, I must approach each nest and flush the brooding mother from it. The more accurate I wanted my data to be, the more frequently I would have to chase her in order to know precisely when various events took place. But my assiduity would be paralleled by an increasing error in incubation time, occasioned by the lapses in duty I forced on the bird. In other words, it was a theoretical impossibility for a human being to determine an accurate incubation period simply because he *was* human. One would have to settle for an approximation, the best one being obtained when the least possible amount of interference had been practiced.

Then the ordinary customs of the bird had something to do with it, too. Seaside Sparrows normally lay four eggs in a clutch, the eggs presumably appearing on successive days. If they began to incubate as soon as the first egg was laid, this one would hatch three days ahead of the last one. Growth of passerine chicks being extremely rapid, these babies would be strung out in an obvious series comprised of a bully at one end and a runt at the other. The result would be that one or more of the younger chicks would be out-gobbled by their larger siblings at feeding time, and very few nests indeed would ever be launching sites for all four offspring. That would surely be an inefficient system. It would also put an added burden on the parents, for part of the energy they had put into egg production and incubation would have been wasted and might have to be made up by their attempting to rear additional broods.

Suppose, then, that nature has arranged things somewhat more happily by giving the female Seaside Sparrow the instinct to defer incubation until she becomes conscious of having completed laying her clutch. The older eggs have meanwhile lain in the nest uncovered. They are nevertheless living objects, poikilothermic to be sure, but endowed with inexorable metabolic drives that distinguish them from inert material. Each egg was fertilized while it was still high in the maternal oviduct—a necessary consequence of the united germ cells' lying on the surface of a yolk mass that must imminently be enclosed, first in a pool of albumen, then with a membraneous envelope, and finally by a rigid shell, all these layers becoming successively less pervious to spermatazoa. Stimulated by the warmth of the mother's body, cell division began immediately, so that by the time each egg completed its structural development and was laid, it already contained a multicellular germinal disc on the yolk surface (though that object could be seen only with a microscope after suitable staining). As soon as the egg was exposed to air a few inches above marshy mud, its temperature would drop and cell division would become retarded. *But it would not stop entirely.* Therefore if the sparrow did not begin to incubate until after the last egg was laid, she would still be sitting on four eggs in four different stages of growth. It is easy to realize, however, that these would be closer of an age than in our opposite case. Hatching of these four should be spread over a much shorter time.

Perhaps an intermediate condition is the natural one, the tendency being toward delay in the onset of incubation. But it is likely that if unusually cold weather intervenes while the clutch is being assembled, the mother will give eggs some protection by warming them to some extent. And as she comes to lay each new egg, she can't help at least briefly incubating those already there.

So where do we stand now? The upshot is that no matter what pattern the mother follows, the minimum natural incubation period will be the time it takes for her *last* egg to hatch. It is not necessary to mark the eggs as they appear in order to know when you are looking at the last one. You assume first of all that the other eggs all had a head start, no matter whether they were or were not officially incubated in advance, and that they will therefore hatch first. The two points in time that you must determine, then, are the instant that the nest finally contains its full clutch of eggs and that other instant when every one of the eggs has finally hatched. That interval should pertain to the last egg's embryology.

An additional important problem still remains to be pondered. When one is dealing with incubation periods as short as eleven days—as Dr. Hill reported for the Sharp-tailed Sparrow

—a difference of one day either way is relatively enormous and one would be greatly in error if he made a mistake of that magnitude. A single day represents 9 percent of the total period in the Sharp-tail's case. Half a day would account for 4.5 percent, and just to bring it to the verge of the ridiculous, 1 percent of the time would be equivalent to two hours and forty minutes of incubation. Yet that is not as ridiculous as you might think when you remember that errors as great as 1 percent are absolutely intolerable in some other kinds of scientific work. Imagine allowing that much leeway in moon shots! Our astronauts would never have made it there and back with that kind of sloppiness.

I must therefore learn the timing of oviposition. Assume that I visited two nests punctually every day at nine o'clock. In one nest the bird laid each egg at eight o'clock and in the other the bird waited until ten o'clock. My records would show that eggs in the first nest took a day longer to hatch than those in the second, though in actuality the eggs were only two hours apart in age. On the basis of an eleven-day incubation period, I would have made an error of 9 percent for the second set of eggs by claiming too brief a period for their incubation.

To avoid that kind of mistake, I would have to study a number of nests, first to establish whether eggs are laid at about the same time of day by all Seaside Sparrows and, if not, what pattern they do follow. Those nests would be so greatly disturbed by the amount of prying I must engage in that I could not possibly use them in the definitive study itself. Consequently after that I must find enough extra nests to give me the assurance of a consistent answer.

Notice that I did not bring up the possibility that eggs are laid at night. That is not usual among birds, and I would wait to see if any hint of such behavior appeared. Likewise I refused to worry in advance about my end point—the hatching of the last chick. Presumably, or at least hopefully, there would be no difficulty in pinpointing that event. Indeed, if egg laying or hatching or or both took place in darkness, I would not be too far off the mark in my estimations, because the short summer nights would separate my dusk and dawn observations by no more than eight hours at the most.

Another lapse you may have spotted is that I have not proposed making any of these studies from a blind or hide. What
better than to work from such concealment right alongside the
nest, with no need for colored bands, let alone telescopes and
binoculars? Perhaps I should have tried that method. But for one
thing, I knew that my actions and the strange object would at
once attract public attention—the sand road was open to everyone, after all—and a curious audience (possibly even a hostile
one) I could do without. Secondly, I was not willing to restrict
my observations to just one or two nests, which would be all I
could study in a given breeding season by such sedentary means.
The results might be sufficiently definite for those nests, but
would they be typical or representative of the species as a whole?
If I could roam around and keep a maximum quantity of nests
under surveillance, my answers might be more valid. Anyhow, if
the rate of nest destruction turned out to be high, I might almost
surely be defeated, were I to study the smaller number.

I measured the marsh in part by reading my car's odometer
along the sand road and otherwise by pacing off various trajec-
tories to points where the land ended at channel edges. The area
that seemed most feasible to work within the boundaries I set
presented a bilobed shape, with a narrow watery inlet almost
cutting it into two separate pieces. I judged the nearer section
to occupy one and one-third acres and the farther one to be exactly twice that large. Since an open marsh has scarcely any permanent landmarks by which to outline a sparrow's usual haunts,
I decided it would be a sufficient test of their homing instincts if
I could find the same birds habitually in either the smaller portion, which I named Marsh A, or in the larger Marsh B.

I had fully intended to settle the incubation question in
1969. Perhaps I have some right to plead that nature played
against my desires, but that cannot wholly excuse the truth that
I became birdbanding crazy and did not concentrate on nests as
diligently as I could have done. But how was I to react when my
very first netting effort of the season, on May 19, rewarded me
with one of my original Seaside Sparrows banded *two* years previously? And when the second session on the next day turned up
another two-year Seaside as well as a Sharp-tail from last year

with one of its yellow plastic bands still intact? I was not going to ignore anything *that* enticing and concrete for the pursuit of nesting data which might elude me in any case. So I may as well tell you my banding story first.

Not that I gave up a search for nests or their close observation, once I knew where they lay hidden. In fact I undoubtedly found more nests than I would have if I were not so keen to net every sparrow in marshes A and B. But this very zeal must have kept the incubating birds in more prolonged agitation and turmoil than is ideal for the expression of natural incubation periods. I realized from the beginning that I might be undermining one study in promoting the other, but there seemed to be no way of avoiding such compromise. I tried to minimize the effects of my intrusion by setting up the nets alternately in the two marshes on successive visits, and also by moving their locations on each occasion by the greatest possible distance. Even so, I would soon find myself plodding through the marsh grass in an effort to drive sparrows into the snare, and those treks—all converging on the target from different directions—invariably led me to set incubating birds into flight repeatedly and everywhere.

Ultimately I had enough sense to realize several facts that made another strategy not only possible but actually better. The centers of both marshes were largely empty of sparrows, most nests being located near their margins. Of course I could not tell where the birds ran about under the grass, but they *did* appear occasionally at borders of a marsh as they fed on open mud, and when they took wing before me they as often flew outward to the shelter of tall reeds, flanking the sand road, as into deeper marshy interiors. Another consideration was my trapping procedure. For my normal "set" I had erected three nets in a continuous row, using four aluminum poles to support them. Each net was 12 meters long, so that after I translated this into English I could say that the nets now spanned about 120 feet. That was so short a distance that sparrows did not have to deviate far from the line I had tried to select for them to miss the row at one end or the other. Or, if they were properly headed for the set but then became aware of it at the last minute, they could often stick to their reluctance to fly high by darting around rather than over it.

A handy feature of marsh netting is that you can easily push a pole into the sandy mud anywhere you choose. For my three-net rig I had prepared four special poles that could be stored in halves in the trunk compartment of the car. I kept nets, bands, and other necessary paraphernalia in the same place, so that I was prepared to do a bit of sparrow banding whenever the spirit so moved me. But now I realized that a mere four poles severely narrowed my scope. On June 8 I bought a roof rack for the car and on the ninth I loaded up every pole from my woodland banding program at the farm. That gave me fifteen additional supports, so that I was able to *rim* Marsh A with eighteen nets around three-quarters of its perimeter. I scarcely needed to enter the marsh to do this, and once the set was complete I had to disturb the birds only a couple of times do get them to entangle themselves somewhere along the line. Indeed I ran into trouble getting later nets up because of having to remove birds from some of the first ones in position. On the next day I ascertained that the same number of nets would go half-way around Marsh B.

This innovation may not have solved all my problems, and it certainly demanded a lot more work, for henceforth it took me well over an hour to get everything ready and sometimes longer than that to close shop in the evening as sparrows made a few last-minute feeding sorties at dusk. But even if the nets continued to miss a few sparrows in the marshes, those must have been exceptionally wary birds that I never would have caught by any method. The chief novelty now was that I quickly banded most of the resident population and thereafter kept renetting many of the same birds. I no longer bothered with colored plastic bands, but did keep a close tally not only on the occurrence of a bird in Marsh A or Marsh B but even the very net in which it was taken.

The thoroughness of the new system showed up in my record of returning birds from previous years—and this is what kept me agog from the banding standpoint and prevented my making the more pedestrain study of incubation time. The fact that I had recorded no "returning" 1967 birds in 1968 must have been the result of my sketchy 1968 attentiveness to the marsh, rather than to the absence of such sparrows, because of the eight Seasides banded in the first season, *four* reappeared two years later. When you consider that a common figure for the mortality rate of small birds such as sparrows is 50 percent per year, that record becomes doubly remarkable. In the first place, one might have expected only two surviving birds in 1969. Secondly, the finding of four survivors at Stone Harbor suggests a return rate of 100 percent for the species; had the eight birds miraculously survived, *all* might have come back to this tiny area.

Those unlikely figures were shown to be atypical when compared with the scores for 1968 birds. Though I had banded fifteen Seasides in that year, only three (with their pretty plastic bracelets) came back. However, that was still a respectable 20 percent rate. Perhaps a combination of the two groups, since each was small, would yield a fairer measure of how Seaside Sparrows normally perform. In that case the seven returnees out of the two-year total of twenty-three birds represented 30.4 percent of the population. And *that* was something new since the days of Dr. Glen E. Woolfenden at Lavallette! Now in possession of sex lore that had been unknown to me when I turned from sandpipers to sparrows along the sand road, I could make retrospective diagnoses of the returned birds on the basis of their current cloacal protuberances and brood patches. This disclosed four females and three males in the group. Thus the sexes were equally faithful to the area—an additional item of ornithological knowledge.

During the first year I had not discriminated between marshes A and B, nor had I made a record of net positions in my notes. Therefore I cannot say anything about pinpoint performance for the four 1967 returnees. However, the three from 1968 appear to have made perfect landings. Two of them, banded in Marsh A, returned to Marsh A. The third one was originally netted on the border joining A and B, so that I recorded it as A–B. Thus whatever it did would qualify as correct. But it was possible to check this bird within the ensuing nesting season. If it had flown into nets on both sides of the line, I could not have given it credit for possessing homing pigeon instincts. But, having reappeared first in Marsh A, it continued to reside there.

In fact I found all of my Seaside Sparrows close to their chosen premises until well into the nesting season. After young birds had fledged, some of the adults began to move more widely. But prior to that time—up to the middle of June—I encountered not a single roving bird. Among those which I succeeded in netting more than once, ten males, seven females and one bird of unknown sex lived in Marsh A, while four males and four females adhered to Marsh B. (Those accounted for half the birds I had banded. The others, which thus had been netted only once, were either transients or else they became net-shy after a single man-handling experience, a rather doubtful accomplishment.) One can see that Marsh A was more popular than Marsh B, despite its smaller size. I believe that A was an inch or two higher than B, and was consequently less subject to tidal inundation, though the sparrows could not have been aware of that safety factor. Possibly they saw something in the resultant character of the vegetation that advised them of its greater desirability.

The 1970 returns would amplify these results. My method of perimeter netting had enabled me to catch thirty-four new Seasides in 1969. These, with the seven returnees, gave me a known living population of forty-one birds and the mathematical expectation (at the 30.4 per cent return rate) of seeing twelve of them again following their spring arrival from the south. Nevertheless I doubted that new patterns would become manifest. It would now be the documentation of duplication that became mere routine, while incubation studies at last demanded breath-holding.

But let me not slight the Sharp-tails. Comparative information for those birds accrued in much the same way as for Seasides, though in this case I had already been scooped by Bertram G. Murray, Jr., of University College, Michigan State University. As reported by him in *The Auk* for April 1969, he banded four Sharpies at the Lower Souris National Wildlife Refuge in North Dakota, and two of these returned the following year. My data did not reflect Mr. Murray's 50 percent rate, but they fortified his finding that the Sharp-tailed Sparrow has a strong tendency to return to the same breeding area in successive summers. My solitary 1967 bird had not showed up again, but of six 1968 sparrows, two returned—one male and one female.

Significantly, Mr. Murray had banded an excess of males, and the fact that I did, too, means that we *both* confirmed one of Glen Woolfenden's discoveries. He had found that male Sharp-tails don't help out at home at all, even after the eggs have hatched, but spend all their time roaming in search of receptive females. Hence they cover more ground than male Seasides do and are likewise netted with higher frequency than the females of their kind. My six 1968 Sharpies included five males, while the proportion in 1969 was nine males to one female. The females —one from each year—were both associated with Marsh A. Of three males that I succeeded in catching more than once, at least one moved between A and B; a second got as far as A–B, while the third was taken both times in A.

A further difference between the species at Stone Harbor was that Seaside Sparrows disappeared from both marshes after the end of July, whereas the population of Sharpies showed a great spurt in August. I could not tell whether these were all birds of the year or whether some adults—no longer with external evidence of their sex—were present. I banded about three dozen such sparrows. Would this give me a great surge of returns in 1970, or did August's contingent have nothing to do with the resident breeding population? Neither Woolfenden nor Murray could answer that question, and I felt hesitant about making a prediction on my own. Still, if we were going to have a 1970 sweepstakes and I was already committed to twelve Seasides, why not indulge in a Sharp-tail bet? Not counting the August influx of unsexed birds, I had records of a dozen Sharpies, new and old, during 1969, one-third of which would equal four. So that's where I rather falteringly put my money.

As I have already intimated, my incubation studies did not come off properly in 1969. I have suggested further that natural adversities and personal inattentiveness were responsible. It so happened that natural adversities were so much more destructive than I would have imagined them to be beforehand, that a full quota of personal attentiveness might still not have given me the information I wanted. The histories of some of the eighteen nests I found will disclose the various difficulties I met.

Nest Number 1, discovered in Marsh B on May 23, already

stand still at times, but soon surging ahead for a few more inches. Abandoning my nets for the moment, I would make a quick round of the nests to check on their welfare. The surface of the whole marsh was already under water, though you would not see it unless you were standing there. How the tide had come in so swiftly and yet so softly was a marvel. All the nests were set on grass stilts, their framework woven into a set of upright stalks, so that at this preliminary stage they were still an inch or two above the flood and comfortably dry.

Returning to the edge of the marsh, I could sometimes not be certain whether or not the tide had advanced further, so I would mark the present waterline with a calm shell. If this were to be a favorable time, the shell might never become an islet, and next day I could see it still resting on the thin line of jetsam left by the farthest foray of that threatening skim of sea water. It seemed that those episodes often occurred in late afternoon or early evening, so that I must leave before the tide turned. To know that water had just reached the lowest nests—that perhaps it was touching a set of eggs or that a brood of young nestlings were dipping their potbellies into it—these were not soothing night thoughts after I went to bed. The outcomes were predictably variable. Sometimes all nests remained safe. Or else they were totally washed out. In intermediate cases, some nests survived while others were lost, or parhaps a particular nest would suffer partial damage by losing a fraction of its eggs or young, the remaining ones seeming not to have been injured by whatever exposure, such as partial immersion, they must have experienced.

The worst times for me were when the tide came racing in and I was there to see it. There would then be no question of marking watery incursions with calm shells. July 24 was such a day. By chance I found the last nest of the season on that date. This was Number 18, situated in Marsh B and containing three eggs. I had discovered the nest at 4:30 P.M. Only one-half hour later I saw those eggs float away, the nest being completely submerged at 5:00 P.M. That must be the shortest ornithological history ever made.

Naturally I had not yet become very strongly attached to

those eggs. What really caused me to rail against the insensitivity of physical nature was the simultaneous plight of nests 16 and 17. I had been watching those since they contained eggs and was now following the development of their broods. The babies in Nest 16 were smaller and simply disappeared during the few moments while I was wringing my hands at the other nest. Seventeen had started out with three eggs, only two of which hatched, and one of those nestlings had already disappeared from some unknown cause. I was now witness to the drowning of the single survivor. This nest had a beautifully woven open-fretted canopy over it. As water rose to cover even that dome, the baby floated upward until it lodged in the inverted cup. There it made a few feeble swimming strokes with its unfeathered wing tabs and then became inert flotsam.

When I forget several rough spots that had me almost gibbering, I can say that my 1970 program came off perfectly. I won both bets about returning sparrows by a needlessly wide margin, and I learned the Seaside Sparrow's incubation period —well, I think I did.

My second nest of the year (Number 20 in the grand sequence), gave me all the necessary information. I found it on May 27 while it was still empty, and then ascertained the probable hour of egg laying as follows. The first egg, present at 3:30 PM. on May 29, was still solitary that day at 8:30, fifteen minutes after sundown. I could therefore conclude that the bird did not lay in the late afternoon or early evening (provided it followed a regular twenty-four-hour cycle of rhythmic periodicity, as is commonly true among small birds). At 9:20 A.M. the following day there were two eggs. Thus the mother could have laid during the night or else early in the morning. Two days later there were three eggs in the nest at 6:40 A.M., and four that evening. Splendid! She must therefore have laid the last egg shortly after my morning visit, i.e., between 7:00 and 9:00 A.M. Now I had a good starting point for timing emergence of the future babies.

Hatching took place on June 12 and 13. On the twelfth at 8:15 A.M. the nest held four eggs, but at 7:30 that evening two babies had emerged from their shells. The next morning at 6:30 A.M. I found three babies and one egg. At 11:50 A.M. the last egg was still unhatched, and I began to fear that it was sterile. However, at 4:50 P.M. the head count was four. Elapsed time since appearance of the fourth egg was therefore twelve days and some-odd hours—definitely less than twelve and a half days, so that one might readily call it twelve and a quarter, unless such precision is more than the data warrant.

Anyhow, that result, although exceedingly satisfactory, cried for corroboration. What if other nests gave widely different results? Had my bird been careless or inattentive, so that the developmental rate of her eggs did not represent the average for Seaside Sparrows? I could not believe such a possibility, but it was nevertheless important to duplicate the twelve-day determination at least once, and preferably at several additional nests.

But now hold everything! Who is this Barbara Blanchard De Wolfe, and what is the heresy she speaks? One evening I was wallowing in sparrow lore in those precious three final volumes of Bent's *Life Histories,* when I came upon this unbelievable passage in Dr. De Wolf's account of Nuttall's White-crowned Sparrows in California: "The eggs . . . do not necessarily hatch in the same order in which they were laid." Just like that, as if this were a perfectly ordinary detail not needing further comment. Of course I was instantly cut right down to the ground, for in such case I had no anchoring point at either the end or the beginning of incubation; this could mean only that eggs varied in their inherent rates of development. Chaos must prevail. If the last egg to hatch in my Seaside Sparrow's nest had not been the last one laid, then that last laid one must have hatched *sooner* and perhaps the incubation period was some fraction of a day *under* twelve days. Must I devise some method of marking the eggs after all, and begin all over again? Perhaps I should emulate Dr. Hill of Barnstable and do away with presentation of the actual details of my study. Did not my present data comfortably "indicate" an incubation period of slightly over twelve days if I overlooked De Wolfe's monkey wrench?

Happily, I knew another lady ornithologist to whom I could appeal in my distress. Mrs. Margaret Morse Nice of Chicago has long been the American arbiter of incubation periods, and if anyone could rescue me, she could. She answered:

"Mrs. De Wolfe's statement that the last laid egg in a set does not always hatch last has been reported many times. I suppose one egg might have been pushed to one side and therefore not incubated as faithfully as the rest."

So much for my ignorance about those "many" reports! But hurrah for Mrs. Nice's information! This meant that variable results, should I now obtain them, need not upset me too greatly. If I could clock events in a series of nests, most of the readings should converge at a minimum figure, with an erratic prolongation only when an occasional egg got "pushed to one side" and received less than the usual quota of maternal heat. While irregularities might remain, they would still follow orderly rules. Chaos had been dispelled.

The trouble was that from now on all my nests reverted to last year's pattern. Sometimes I failed to determine when the last egg was laid. At other times one annoying egg in a set would refuse to hatch. Most often the nests were simply failures, being washed out by high tides or—worse—being destroyed by marauders of some sort. The reason I say "worse" is because I often felt that the despoiler had been none other than myself. I could never be sure about this, but the evidence against me was uncomfortably strong.

I had quickly undertaken total netting of the little marshes, once the question of incubation periods had received its first completed datum. That meant not only setting up the peripheral line of nets around each marsh, but then a series of careful drives from within, to get recalcitrant birds to leave the grassy cover and entangle themselves. Indeed I *was* careful, peering into the greenery before setting down each lifted foot, for in addition to my natural wish to avoid damaging nests, I still wanted new ones on my roster for further observations on laying and hatching dates. During a late afternoon and early evening of netting I would tramp across one marsh or the other half a dozen times. Anyone watching from a distance might have likened my deliberate, hesitant-footed gait to that of a monstrous stalking heron.

What would *you* think, if you came upon a nest with two eggs inside it, another egg balanced on its lip, and a fourth one lying in the mud just below? Wouldn't that look like the result of a heavy blow of some sort, such as a kick by a person's foot? That's what it looked like to *me*. Well, I ran into that tableau, or close variations thereof, more often than I like to admit, though I must be honest and do so anyhow. Nests 21, 22, 27, 31, and 36 all seemed to have suffered the same kind of trauma, and they all lay close to tracks that I had left in the grass while walking across the marshes. I tried to persuade myself that someone else was responsible. After all, I spent only two or three hours here every few days, whereas Stone Harbor was a resort area and there might be hordes of people, especially on weekends, trudging through these marshes heedless of nesting sparrows. But I did not find such ruminations at all convincing.

It was easier to find partial excuses for myself on more direct grounds. I have mentioned that when the fledgling in Nest 17 drowned, it floated up until it became stuck in the woven grass canopy over the nest. That had been a rather open type of fretwork for me to have been able to see through it. My kicked 1970 nests were all topped by solid roofs, under which eggs lay completely hidden. Some unkicked nests were equally well concealed, but I began finding those when I learned to search not only for nests as such but also for telltale domes of basketwork. Glen Woolfenden, by the way, said that stems were not woven over his nests at Lavallette, but those Seaside Sparrows were nesting in clumps of rushes—so-called black grass—that may have been too rigid to work. Cordgrass at Stone Harbor, *Spartina alterniflora,* had a yielding texture, and the flattened—rather than rounded—blades lent themselves readily to crisscrossing and threading.

Any nest that I kicked, then, had really been invisible from almost all vantage points except a direct approach to its entrance hole on one side. I might even have kicked the same grass tussock more than once, before I knew a nest was there, if I happened to repeat my passage always in the same direction. Of course once I found it, the nest was henceforth marked by a nearby stake of some sort, and its days of jeopardy from my great clumping boots were over. Should eggs still spill out, I could then invoke weekenders or other disturbers of the peace and absolve myself from guilt. However, the nests enjoyed peaceful careers.

The closest I came to corroborating my measurement of Nest 20's incubation time was at Nest 37. This one, too, had a well-woven grass top when I found it at 1:30 P.M. on July 3, and since it contained only two eggs, I automatically looked on the mud underneath it for derelicts that might have been dislodged from it. I found none. Moreover, the nest proved itself a new one by holding a full set of four eggs at 11:50 A.M. two days later. That fit the concept of early morning laying, so I now predicted a hatching sequence on July 16 and 17, with arrival of the fourth fledgling some time during the afternoon of the latter date.

How nicely everything went up to that final tiny event! At 9:15 A.M. on July 16 the nest contained one baby, one egg just hatching, and two inert eggs. At 7:30 P.M. that evening there were three babies and one egg. Now my hopes all depended on that last one. At 8:00 A.M. on July 17 the egg was still there. That was all right; it need not hurry to fortify my prior observation—indeed, hurry would force me to modify my conclusion by shortening the duration of incubation by those few final hours. But the egg never did hatch, and I decided next day that here was an additional example of the infertility trick. But that was not the answer, either. Inside the perfect shell I found a dead but almost fully developed foetus that looked perfect, too. It must have died within a day or two of hatching time, though I could discern no physical cause for such a tragedy. My searching in 1970 revealed only three more nests after this one, all with their full number of eggs when found, so I had to restrict my incubation lore to the solitary case of Nest 20.

Nests, nests! Such variation existed among them that perhaps it was futile to look for constancy in their contents. Could their very situation contribute to slower or more rapid dissipation of heat, as the case might be, and consequently lead to differences in incubation time for which the brooding parent was not responsible?

I have not had the chance to get into the problem that deeply, although there can be no question as to Seaside Sparrows' selecting nest sites that range from greatest concealment to maximum exposure. Glen Woolfenden's Lavallette nests managed to be well hidden in clumps of black grass despite their lack of a canopy, and in respect to their secrecy were similar to my nests with woven tops. However, I found a few nests at Stone Harbor that could really boast total concealment, since they were built under boards or windrows of dead reed stems left behind by the tide. I found those nests only by turning over such detritus in a blind quest for whatever might be hidden there. At the opposite extreme were some nests, built in short grass near the marsh edge, that were as open to the sky as butterplates. One of my friends in the Delaware Valley Ornithological Club told me of a day, many years ago, when he found over a dozen such exposed nests in a newly cut salt-hay meadow on the Delaware Bay side of the peninsula. I conclude that Seaside Sparrows tend to build concealed nests and will use whatever materials are at hand to achieve such concealment, but in the absence of any means to attain privacy they will not forego nesting.

Twelve Seasides and four Sharp-tails had been my projected quotas of returning birds in 1970, so the actual respective totals of twenty and seven almost doubled my joy and satisfaction. Aside from raising percentages of returning performances, the records furnished a few new items of information. A negative one pertained to the color-banded birds: not a single Seaside Sparrow from that 1968 group of fifteen came back, whereas three of the older 1967 birds did. Among the Sharp-tails, a solitary female with a yellow tape on each leg reappeared. Color-banding with plastic tape seemed to have a definitely adverse effect on survival. That could have resulted from the birds' increased conspicuousness, or else from physical constriction of the leg developing from unknown causes long after application of a strip of tape. Probably the returning bird that gave me the greatest pleasure when I read its band number was that original lone Sharp-tailed Sparrow of 1967. For the following two years she had eluded me, but here she was in my hand again, raising her class reunion attendance record from zero to 100 percent. In actual fact, I now figured Seaside Sparrow return rates generally at 42 percent and Sharp-tails at about 15 percent. That made long-

term work with either species a refreshing pastime, for the handling of known birds from past years lent a feeling of continuity to
any project.

And indeed I found myself becoming more and more involved in questions that loomed as projects rather than as simple
games with percentages. All those nests demanded that I identify their owners, especially after high tides when new nests
quickly appeared close to sites of older ones that had been
washed out. Had the original females rebuilt on their former
home grounds, as if failing to learn from disaster? Or had they,
in fact, moved away and left the area open for settlement by
newcomers not yet indoctrinated in the treachery of local tides?
I had had rather poor luck at nest trapping in former years, but
that had discouraged me too quickly. Now I must develop a more
effective technique and insinuate myself into the secrets of
marsh sparrow family life.

During my visit at Penn State, Dave Davis's graduate students had caught nesting Red-winged Blackbirds by setting up
two nets in a V-formation with the nest between them near the
point of the V. A returning bird was likely to fly into one net or
the other from the outside, though there was also a minor possibility of its entering the mouth of the V and gaining its nest. In
the latter case the students, who had been watching through binoculars, would approach the open end of the V, whereupon the
bird would fly into one of the nets from within. Later I read of
an ornithologist who used much the same technique successfully
with nesting Bobolinks in grasslands of Wisconsin. Since both
these cases involved ground-nesting birds in open, flat terrain
devoid of shrubby cover, I assumed that I could apply the
method to my Stone Harbor sparrows.

At first I used a two-netted V, and when that turned out to
be relatively ineffective I tried a square-bottomed U, using three
nets, but with even less success. I even experimented with a net
spread on the surface of the marsh grass over a nest. The Stone
Harbor problem—or rather the Seaside Sparrow problem—as I
eventually concluded, is that this kind of bird does not fly to its
nest. Such a maneuver obviously would reveal a nest's position,
and I suppose that Redwings and Bobolinks haven't much choice

about giving that information away to whatever eyes are avariciously watching their behavior. However, Seaside Sparrows—adapted as they are to foraging on mud in a marsh—invariably walk to their nests in total concealment. Given sufficient advance warning of the approach of a potential enemy, they leave the same way. In both cases they easily passed under the edge of the net. Thus I often had to feel the eggs in an apparently deserted nest to reassure myself that they were in fact still being incubated. At times, on the other hand, a bird will sit very close, on the chance that the intruder will pass by at a sufficiently safe distance. If the threat comes too near, the bird still tries to run a few feet away before rising into flight, but its sudden and obviously agitated appearance remains a banner of what lies close by.

I therefore adapted my netting methods to those facets of the birds' behavior. Using three poles at the corners of an equilateral triangle, I set up three nets so that they completely surrounded a nest at the center of the triangle. I then retired to the car or to the other marsh for half an hour or so. Even though I watched furtively, I saw no sign of a bird's approach to the triangle. Then, advancing in an oblique fashion that was supposed to look as if I would bypass the nest by a goodly margin, I nevertheless managed to get near enough to attempt a surprise strategy. Suddenly I made a dash at the triangle. Often nothing happened; cool eggs told me that the bird had not been there, while warm ones proclaimed both her presence and her discreet departure. When I was lucky, a bird would fly up and get caught. I never felt that it was putting undue demands on the circumstantial evidence to ascribe that bird to that nest.

Not that even this technique worked very well. I failed to catch birds at many nests. After the eggs hatched, and especially when the babies were at their hungriest close to fledging time, both parents attended a nest at frequent intervals. A perfect catch then would give me an enticing basis for keeping tabs on the future behavior of that conjugal pair. But the chances were only slim for a second perfect catch of those same birds. Indeed my records of retrapping single birds were few enough. However, I can report that four 1969 females nested in 1970 not only in the same marsh as before but within 200 feet or less of their respective former nest sites. And one pair that was flooded out in Marsh A early in 1970 renested only 100 feet away in the same marsh almost immediately.

Then of course there were Nests 24 and 26. I spoke of almost gibbering while trying to clock incubation periods. One of those times occurred when I was putting a triangle around that precious Nest 20 which finally gave me my one acceptable answer. As I maneuvered nets and poles to get the nest at exact center, I suddenly became aware that I was beholding *two* nests. Impossible! But there they were, as alike as Tweedledum and Tweedledee, each containing four warm eggs, and a mere (as I later carefully measured it) nine-and-a-half feet apart. These were near the edge of Marsh B, in short grass, and therefore of the open, exposed type with scarcely any concealment beyond a few green stems and blades that happened to grow at their rims.

As I now stared at this spectacle in disbelief and bewilderment, it struck me that perhaps I had been taking my vital notes first at one nest and then at the other. This was where I thought I had verified the time of day at which Seaside Sparrows lay their eggs. What good were my scribblings now? Calming myself as best I could, I tried to remember the exact looks of the first nest when I had found it, to see if I could not persuade myself that I had studied only that one after all. In addition I had a good look at such remnants of my footprints as I could find. At last I *did* decide which was the original Number 20, but that was no proof that I had not confused it on one or more occasions with its neighbor, before becoming aware of this absolutely baffling finding.

There was nothing in Glen Woolfenden's paper to prepare me for such a situation. He observed that male Seaside Sparrows established ample and well-defined nesting territories which they defended against other males. Each territory was occupied by only one female. Could it be that I had discovered a polygamous trait in this species? That must be the answer, unless . . . but of course! I hastily finished erecting the triangular net-set.

In short order I was so fortunate as to have captured two female birds within the triangle, one Seaside and one Sharp-tailed Sparrow. Later, the hatching times in the two nests were sufficiently widely separated to prove that I had not mixed up my notes.

In any event, the discovery of Nest 26, while helping to crack the mystery of the whereabouts of Sharp-tails' nesting at Stone Harbor, simultaneously created a new one. Glen Woolfenden had noted that Sharp-tailed Sparrows at Lavallette nested in drier areas of the marsh than Seasides, but here my birds were living as back-fence neighbors in the identical environment. Nest 24 turned out to belong to another Sharp-tail, both to my astonishment and my delight. I had treated it in my field notes as a Seaside's nest until my triangular net-set caught a female Sharp-tail—my long-lost 1967 bird, in fact. But this nest, too, was in Seaside Sparrow nesting terrain, near the center of Marsh B and close to Seaside nests 27, 28, 36, and 37. On the other hand, I had to conclude that Sharp-tailed Sparrows' nests were definitely rare among the clusters of Seaside nests in my two marshes. And the post-nesting influx of young Sharp-tails each August must have been bred in some different though nearby area. Obviously any terrestrial terrain other than these tide-level marshes *must* be a drier one; so Glen's Lavallette impressions probably applied to Stone Harbor also.

I do, incidentally, take issue with Glen Woolfenden on one very minor point. He concluded that Seaside Sparrows forage only at marsh edges, not within the interior expanses. While I cannot prove that the birds behaved differently at Stone Harbor, I simply feel that they must have. Occasionally during the day I would observe one or two birds engaged in such foraging, but for the most part the marsh would seem empty. Yet, at sundown, a male would suddenly pop up somewhere and begin singing from a grassy tuft (or later from my nest-marking stakes). At this signal another bird would pop up and then another until the full population erupted into view. A great deal of chasing would then take place, and some birds came down to run briefly along exposed sandy or muddy margins of the grassy sea. But my conviction remains, that in order to keep themselves alive—let alone feed nestfuls of rapidly growing babies—the sparrows must have spent many hours during the day in concealed hunting among the grass stems.

The year 1970 had been pretty good. It simply did not last long enough. And there were points in my program that I was now able to criticize in retrospect, though I could remedy them next year. Most important of all: I must try to get to the marshes before the sparrows did in 1971. In the past I had deferred marsh sparrow studies until migrating warblers had finished their rush at the farm. By the time I set my first nets at Stone Harbor, it would be about May 20 and early nesting would already be in progress. I had missed many exciting weeks—unless the birds merely flew in and settled unceremoniously on their old perches (an unthinkable possibility).

I probably ought to begin weekly netting sessions about April 1. That would be a month in advance of the sudden spring-time increase in abundance of seaside Sparrows on Cape May County, and would give me a chance to check on the species' winter contingent. It has long been known that some Seasides can be found at this latitude all year long, but formerly the winter birds were thought of as only a sprinkling of diehards. However, the tally of Seaside Sparrows in the Cape May area as noted by observers participating in the 1969 Audubon Christmas bird count was seventeen. Seventeen would constitute a big fraction of the Stone Harbor summer population if those birds were all wearing my bands.

Furthermore I must find out whether males return first to my little marshes A and B, to be followed by females after the males have settled their territorial disputes. That is the way many songbirds manage things. But I have a feeling that Seaside Sparrows may follow a different custom. If the birds mate for life (which I doubt, because that is not the way of other sparrows), one might expect them to arrive at the marshes simultaneously, for it would be natural for couples to remain together through-out the year. That would accord with the fact that my females returned as faithfully to their respective former marshes—A or B, as the case might be—as males. But if males arrive first, and if there is no permanent pair bond between the sexes of Seaside Sparrows, one might expect the later arriving females to show a less constant attachment to the particular marshes they once had occupied. Mind you, their returning instinct might be just as strong as the males', but that would be more likely to apply to the area in general than to specific nesting territories, and females might then be found scattered randomly in Marsh A or Marsh B without any relation to their erstwhile whereabouts.

But the contrary could be true. If Seaside Sparrows are evolving in the same direction as Sharp-tails, in which the male has given up domestic responsibilities, perhaps the female Seaside *does* have strong territorial passions. It would be extremely novel to discover in 1971 that female Seaside Sparrows arrive at Stone Harbor before the males and that it is they—the wives—who settle boundary lines for the forthcoming nesting season.

Oh, those mighty sparrows! Come to think of it, I guess I better write to the Fish and Wildlife Service for a permit to band nestlings in addition to adult birds. Are missing oldsters replaced in marshes A and B from local cradles? How else could one find out? Banding babies is frowned on in general, because of bands being wasted on youngsters with notoriously high mortality rates. But in special circumstances the Fish and Wildlife people are willing to make exceptions, and I suppose that by this time I can present a fairly good case.

Chapter 8

MICE AT THE TABLE

What puzzled me was ragged holes in the paper towel where I laid bacon strips each morning to absorb excess grease. Living in semicamp-style at the farm, I did not bother to use a fresh piece of toweling every day. Sometimes the same sheet stayed on the enamel surface of the stove for a week. By that time it was so soggy with drippings that my bacon would no longer get crisp and dry the way I like it. But how did the holes get there?

As I sat munching shredded wheat, musing over the ragged mystery, I sometimes had the feeling of being not alone—of being spied upon. Then, looking up quickly, I would imagine that I had seen a small brown shape flitting behind a pan on a rear burner of the stove. Gradually it came to me that I had mice. It must have come to the mice simultaneously that I was innocuous, for in time to my increasing curiosity their timidity dwindled. Soon they could hardly wait for me to finish cooking my bacon. As soon as I had sat down to breakfast at the kitchen table, they popped into view, from a canyon of pipes and wires behind the stove, and even up through burners that I had not used for bacon or coffee and which were therefore cool enough to use as climb-ways.

There were two of them. Out they came and deployed themselves as if on a highly trained schedule, one to each of the forks I had used for turning bacon rashers. Big ears, big eyes, long tails, brown backs and white bellies proclaimed these *Peromyscus leucopus*, white-footed mice. Now I could add "big tongue" to the description, as they licked the four tines successively of their delicious fatty residues. Only when that rich source was de-pleted did they begin to eat the paper. Poor things! they seemed unable to extract the goodness and spit out the pulp, so paper became inseparable to their sandwich. So much the better, then, that my housekeeping was done by the week instead of daily. Mondays, on the eve of ravages by Mrs. Fisher, my cleaning lady, were the richest. Then mice came nearest to a diet uncomplicated by roughage.

This was my first winter at the farm. The previous owners had latterly kept a dog but no other livestock. I had not yet acquired pigeons, nor had I established a permanent birdbanding area near the house. Therefore the premises afforded poor hunt-ing to rodents. Those connoisseurs of man's untidy wastefulness, the house mouse and the Norway rat, may have wandered near with noses twitching inquisitively, but they did not perceive the right odors—or did not judge aromas to be strong enough—to make an invasion interesting. Later, when my lavish strewing of pigeon feed and wild bird seed became known to all the gnawing creatures in Eldora, I would be haunted by them even into my bed at night. White-footed mice could not compete with that brassy lot. But this winter the house was theirs, and I cooked ex-tra slices of bacon on Wednesday morning to give their new paper towel a good start toward saturation.

Lovely though I found these pets, I now could blame them for many dark pellets that had appeared on the stove and among unwashed dishes on the shelf beside my kitchen sink. I noticed also that they made a veritable banquet area of the frying pan after I set it aside to cool each day with its morning quota of congealed fat. Footprints, tail tracings, tooth-gouged gulleys— and further pellets—testified to the treat I had left for those lucky mice. However, they were just a bit too lucky, or too on-coming, to suit me. This degree of intimacy could not be very sanitary. I vaguely went over the list of diseases shared by men and mice and dragged up such items as dwarf tapeworm infection and lymphocytic choriomeningitis. Much though I hated myself for the decision, I ordained that the house should be my unshared habitat and that intruding baconlovers must go.

Besides, midsummer would be approaching. I had always heard that even the highly domesticated house mice will leave farm residences during the hot months and revert temporarily to a feral life while it is easy for them to forage in nearby fields and hedgerows. Then white-footed mice, being woodland creatures, certainly had no right to remain lingering behind my stove in July. My house stands entirely in the open, with the closest woods lying across a busy highway, so it was not a question of these mice merely visiting me for a breakfast snack and afterward spending the rest of the day in hidden runways beneath the forest litter. No, they were definitely permanent boarders and lodgers, so warped by my hospitality that they had eschewed the normal ways of their kind.

If I had had a magic wand, a single wave of which would have eliminated all mice currently present, I suppose I would have waved it without hesitation. But since I must use some practical method of extermination, I was driven to preliminary thought in the matter. And out of that delay grew even further reluctance to engage in the project. I discarded the idea of poisoning my mice immediately, not only because that seemed a nasty thing to do to them, but also because I did not like the idea of having poisoned carcasses about, either inside or outside the house. Likewise I was repelled by the thought of snap traps, because that method sometimes injures victims without killing them, while of course if the mice *were* killed, they would be just too dead to suit me.

The obvious compromise was to catch mice alive and then release them in some faraway woods whence they could never find their way back. Once I was satisfied with that decision, I sent an order immediately to H. B. Sherman of De Land, Florida, for a dozen of his collapsible aluminum live traps (mouse size), and I also bought locally a couple of Size O Havahart traps. It passed through my mind that I should make up a batch of the mammalogist's classical bait: oatmeal, corn meal, peanut butter, bacon fat, and raisins. But that seemed more elaborate than necessary for my simple problem, and I decided to make do with plain peanut butter.

Naturally the mice walked right in and trapped themselves. Why shouldn't they? There was nothing to fear in this house—at least there never had been. The heavenly aroma of peanut butter must convey only another token of the generosity and benevolence of the resident hominid. Anyhow, I had set the two Hava-harts on the stove, and now I had two mice. The question was, "How far away should I banish them?"

My woods on the side of the house away from the highway had to be reached by crossing a wide expanse of lawn. After that, one must traverse a vegetable garden or an orchard, in two of the three possible directions. The third access led directly into an immature swampy forest overgrown with catbrier. This last area was closest to the house and may well have been the source of my mice. However, I decided to make the challenge a bit tougher. The route across the orchard to a path entering one corner of my mature and relatively dry woods was 116 yards from the back door, and I felt that any mouse venturing—and then actually managing—to return from that remote otherworld would have to be a very enterprising mouse indeed.

That was on July 21. On the twenty-second I caught two more mice on the stove. So I had had more than two all the time, but they had not all appeared together! I released this pair at the same corner of the woods. On July 23 I caught a third pair, but after that the number dwindled to one daily. At least I thought so. Sometimes I would catch two on a given day, but one would enter the trap at night and I would find it when I came down to breakfast, while a second would appear in the afternoon. But by September 22 I had liberated fifty-three white-footed mice, and that simply seemed a most doubtful fact despite my having kept an impeccable record of each trapping and release.

How could I be sure that these had all been different mice? I remembered a rule of thumb taught to me by pest exterminators when I was working with domestic and wild rodents on a murine typhus fever problem in Tampa, Florida. "However many rats you see, multiply that number by twenty-five to establish the actual numbers present." My farm data thus indicated that I had been serving bacon drippings to a population of 1,325 deer mice. Ridiculous! There wasn't that much room behind the stove.

Obviously the rule of twenty-five did not apply to the kind of creature I was dealing with. But equally clear to see was that some of the mice had been finding their way back from the woods. My next task must be to learn how many of the individuals were able to perform that feat and how long it took them to do it.

While working with small mammals not only in Tampa but also in various tropical regions of the world, I had repeatedly found that these creatures get along perfectly well minus a toe or two. It is therefore possible to give them (under anesthesia, of course) permanent ID cards by clipping their toes in definite sequences and patterns. Rodents have no thumbs, so that on the forefeet one can remove individual toe number 2, 3, 4, or 5 on the right or left foot of the first eight animals, and then single hind toes for the next ten captives. Number 19 would be the first animal to have to lose two toes (left front 2 and right front 2, for example), though I hoped not to be confronted with surgery that extensive in my kitchen.

I quickly discovered that I had been dealing all along chiefly with *one* mouse. FL-2 (for front left toe Number 2 missing) was an adult female that I marked at 1:15 P.M. on September 22. After that her captures were less traumatic, but one would imagine that it would nevertheless remain harrowing to a small mouse to hear the jarring crash of front and back doors of the Havahart trap, then to languish for an indefinite time in the trap until the hominid appeared, forthwith to be chased by a blast of horrible breath from the trap into a cloth sack, thereupon to be grasped through the cloth until one's life was almost squeezed out while one's feet were examined, and finally to be transported for a long, jogging journey in the sack to those alien woods at the other side of infinity. But in the next few days she reappeared in the trap on the stove notwithstanding as follows:

September 23. 7:00 P.M. and 9:35 P.M.
September 24. 9:30 A.M. and 7:30 P.M.
September 25. Present at breakfast and at 8:30 P.M.
September 26. Present at breakfast.
September 27. Present at breakfast, at 8:30 P.M. and at 9:30 P.M.

At the time of her first recapture on September 23, the stump of her missing toe had a scab on it and must still have been slightly sore. That did not seem to inconvenience her in covering long distances or to deter her from reentering a danger zone. *Ergo* she was stupid or else her definition of danger was different from mine. Anyhow, it seemed to me that she had made two points—that she lived in my house and that 116 yards were a mere nothing to a white-footed mouse.

When FL-2 made her expected reappearance in the trap on the morning of September 28, I decided to see if I could learn something of her technique. Whenever I had released her in the woods, she had always darted away immediately to hide in the nearest leaf litter, so I had not been able to watch her return progress. On this day, however, she must practically have followed me back through the orchard, for she was in the trap once more within twenty minutes of her liberation. Since the back door was closed, she had made her way through a well-known opening in the cinderblock foundation, and that was the trick I now wanted to witness. Accordingly, I undid the neck of the sack in the middle of the driveway at a spot which she must have crossed dozens of times. To my amazement she acted as if bewildered and took her way falteringly into the front shrubbery in anything but an exhibition of beeline homing behavior. However, it may have been my presence that distracted her, for not long after I went away she was back in the trap on the stove. I decided to try again. This time she ran from the driveway straight to the side of the house, climbed up a cinderblock and disappeared in a crack where masonry met the clapboard superstructure. I don't know whether or not that was the portal she customarily used. If so, I can't imagine how she discovered it in the first place. But if not, she wouldn't have been able to find it as unerringly as she just had.

Anyhow, that demonstration was enough for me. As a matter of fact, I had other things to do besides releasing the same white-footed mouse over and over again. But since it was now proved that I was not really overrun with mice, I decided to let this one have its way with less disturbance from me. For the time being I put the traps away. Occasionally I saw a mouse, but in between such sightings I still had the comfortable sense of its presence whenever new holes appeared in the fat-saturated paper towel on my stove.

Unfortunately that was a misleading criterion. One morning, only a week later, I thought my vision must be a bit bleary at breakfast, for the mouse not only looked drab but its ears seemed small and its eyes were definitely contracted and squinty. A house mouse! At last they had come. Naturally it was my fault, because by this time I was scattering grain lavishly in my bird-trapping area just outside the door of my tiny laboratory, and no wide-awake, opportunistic house mouse, hitchhiking along the highway, would dream of moving further down the line once it came within olfactory range of that spread. Moreover, there must be plenty of nooks in which to make nests in the crawl-space under the house. Well, then, one might as well come upstairs—or up pipes—to explore those other delicious smells.

Despite my almost catholic enthusiasm for all kinds of wildlife, I have a few mild prejudices. One of these was against house mice if they were to supplant white-footed mice. House mice are human-transported invaders from the eastern hemisphere; they are neither as attractive nor as gentle as our native mice; and they have an offensive body odor. On all those counts I would declare war on house mice if they became too aggressive. But first I must see what sort of arrangement the animals worked out themselves. And while learning about that, I might as well give the house mice a few familiar chores to do about the place. Consequently that particular one, also an adult female, became FL-2 in the *Mus musculus* series and was deposited at the corner of the woods in the spot already intimately known to FL-2 in the *Peromyscus leucopus* series.

One of the traits that has made the house mouse such a worldwide success is its ability to learn quickly. I would probably have kept on catching my white-footed FL-2 for the rest of her life if house mice had not arrived. But now she soon disappeared. I caught her on October 3 and 4 and for the last time on the tenth. Two other white-foots, FL-3 and FL-4, appeared briefly but were absent after the twentieth. And that was the end of white-footed mice on the stove. But meanwhile I continued to have house mice, though each one let itself be trapped only the first time—or at long intervals. They were just as bold about licking bacon fat from tines of the two forks as I sat at the breakfast table; they were simply smart enough to know what was good or bad for them and to act correspondingly. And while it really has nothing to do with the price of eggs, I must indulge in the sentimental observation that the house mice, for all their company and duplication of white-foot antics, were not nearly as cute as their predecessors.

But the house mice *did* make it safely back to the house from the woods. Since they mostly shunned traps after their first costly lesson, I could not time the speed of their return, and I had to rely on modified tactics or on chance alone to renew my contacts with them. Intermittent trapping caught any new house mice that might arrive from time to time, but only rarely would a marked animal be retaken, its memory apparently having become dull. Mostly I found my stump-fingered tenants through accidents they got themselves into by their own exploratory curiosity. One fell into a partly filled bucket of water and drowned. Another jumped into the deep plastic trash container in the kitchen at a time when I had removed the trash bag but neglected to replace it with a new one. That mouse had probably been foraging in the container regularly, but now it was confounded by the smooth cliff walls rimming it in a full circle.

And what an experience house mouse FL-4 had! While I was cooking my bacon one morning, I kept hearing a faint, unfamiliar sound, but I was unable to trace it and actually did not bother very much about it. Then by chance I happened to look into the tin can into which I regularly poured bacon drippings. I had emptied the can recently, so that it held only a couple of days' accumulation of fat. Otherwise FL-4 surely would have drowned, and I might never have discovered her carcass. But as it happened, there was about an inch of semiliquid grease in the bottom of the can and my mouse was utterly and thickly covered with it, with her feet able to touch bottom and her snout and eyes just protruding from the surface.

It is famously difficult to catch a greased pig, though one reads that some people manage to do it. Not knowing where to look up the approved method of handling greased mice, I had to figure this one out myself. Former experience with the empty plastic trash container suggested the first step. Removing the bag, I dumped out the mouse into the container with as little fat as possible. Then I lowered one of the Havahart traps and drove the mouse into it. Now I was able to place the trap on many thicknesses of newspaper on my laboratory table, and to push strips of paper toweling through the wire meshes of the trap. I thought that FL-4 might sense that the shredded paper would make a good nest. If she tried to gather it about herself, it would absorb some of the grease. However, the mouse was so self-absorbed in her own predicament that she paid no attention to the paper or to me. She simply set herself to a grooming program that ought to have sickened her if she were indeed not already deathly ill after her night in a well of bacon fat. How much of the material she had swallowed then—and must lap up now—who can guess? What did all that fat do to her digestive system? What strain did her kidneys withstand in handling all that salt? After a couple of hours her fur looked almost normal. I took her to the woods and opened a door of the trap, whereupon she scuttled away, seemingly unharmed. I never caught her again. Yet she may still have been one of my mouse tenants, the wisest of them all.

Eventually I declared the war I had threatened to wage. All house mice must henceforth be destroyed. I had not suddenly taken more strongly against them, but they were simply becoming too abundant. They would enter my bird traps in broad daylight to eat the expensive wild bird seed I had placed there. They would die in various unreachable places in the house, such as the hot air conduits of my heating system, whence their unbearable effluvium emanated until desiccation and decay slowly reached a physicochemical endpoint. They would appear at any or all times in odd places, like a desk drawer, a book shelf, a window curtain, or a bed. One night I was awakened by being actually bitten on the hand, though the nip did not draw blood. That was just a bit too reminiscent of the days of the Black Death to suit me.

New house mice were easily caught, as I have already suggested. To get rid of the old ones, I resorted to dirty strategy by producing snap traps, devices with which they were unfamiliar. Thus I was able to write *finis* to the history of my very first visitor, FL-2, on December 16, after a lapse of more than two months (though she had reappeared once on October 26). My best longevity record was that of FR-3, an adult male first trapped and marked on November 1 and killed in a snap trap on February 7, after ninety-nine trap-free days of high cholesterol debauchery.

From now on I trapped only in blitzkriegs, using Shermans, Havaharts and snaps when I became aware of new house-mouse invasions. After one population had been destroyed, it might be months before the succeeding one became noticeable or objectionable. During these intervals I stayed alert to the possibility of recolonization by white-footed mice. I could imagine the little woods creatures attempting to achieve sanctuary in my foundations and walls but being held at bay by a ring of hostile house mice around the establishment. If some such force were not operating, why did they stay away?

It seemed not very bright of me to sit and wonder about things that I could easily learn. Thus self-berated, I moved my live traps to the woods. I had seen so many mouse-sized holes in exposed patches of earth there that I visualized the entire litter-covered floor of the woods as being riddled by them. Nothing but a teeming population could have done that much burrowing.

Using Burt and Grossenheider's *A Field Guide to the Mammals*, I had made a list of thirty-nine species occurring in South Jersey according to maps showing the range of each kind. Naturally that included things like bats, opossums, carnivores, deer, and larger rodents, such as muskrats and squirrels, that my traps were too small and frail to handle. But apart from those—and imported invaders—there remained six, or possibly seven, native species that I might hope to catch. As a matter of fact I felt ashamed that the only one I had seen thus far, the white-footed mouse, had had to make the introductory overtures itself, when it was my prerogative as a naturalist to have courted the entire woods fauna long ago. Anyhow, I hoped now to track down white-footed mice to their home ground and to add the rice rat, southern bog lemming, boreal redback vole, meadow vole, pine vole and meadow jumping mouse to my roster of acquaintances.

I was still sticking lazily to plain peanut butter as bait. In a serious mammalogical survey, one would not only offer more sophisticated lures, but one would set a profusion of traps that were, moreover, of more varied sizes and styles than mine. But all I wanted was a rudimentary sample here and there. Even if animals entered at a rate of only one per night, I should eventually piece together a sufficiently good idea of the woods' denizens to satisfy my kind of curiosity. A professional mammalogist must often "trap out" a rodent population completely before he can make an analysis of its components, whereas I would not find that sort of retrospective knowledge very cheering on subsequent woodland walks.

I caught nothing—nor have I on subsequent attempts. Certainly the peanut butter was not wholly at fault, for it had caught white-foots repeatedly in the house. Undoubtedly that bait *was* inadequate for pine voles—I have since read that those short-tailed little creatures are very fussy about menus that mammalogists present to them. But a zero score on white-foots, at least, was puzzling. I rebaited the traps for a few nights (for ants cleaned them out every twenty-four hours) and then quit in discouragement.

Eventually they came back into the house, or so I thought. That is why I purposely am vague and say "they." I heard them usually moving furtively in the wall of the stairway and the ceiling between first and second floors. Often this would be after I had gone to bed, and I sometimes wondered whether a prowler had gotten into the house and was sneaking about in the dark. However, the sound was too light and pattering to be made by a man-sized intruder. But there were definite peculiarities in the situation. For one thing, I saw no mice, nor did traps on the stove catch any. Secondly, the sounds occasionally became too loud for a mouse to have made. There must have been more than one animal in the wall, for after the normal slight scratching noise, a bounding chase from the rear of the house to the front and back again would suddenly take place:

> Scramble, scamper, bump! Scamper, bump, scramble!
> Thump, thump, thump, *thump!*
> Scramble, scamper, bump! Scamper, bump, scramble!
> Thump, thump, thump, *thump!*

Eventually the flying squirrels found their way out of the wall and into the rooms, though I could never find the hole through which they entered. The trouble was that they could not find it again either, and then until I managed to catch each animal in a butterfly net and release it out of doors, it lived an unhappy, foodless and waterless life in my huge house-cage. They hid during the day but tried to fly out of the windows at night, though either glass or screening prevented their escape. My bedroom was a favorite gathering place. Using the bureau as a launching pad, they took off for the bed, landing heavily on my sleeping form, and thence tried to clamber to the windowsill. On my return from a weekend in Swarthmore, I once found a flying squirrel that had drowned in the toilet.

When I began breeding Fantail pigeons, I had Rudd build several cubicles, where I could isolate individual pairs, in an old chicken shed. A highly important part of the construction consisted of rat-proofing with sheet metal flashing and half-inch mesh hardware cloth, for my life had been scarred on several occasions by plagues of rats that slaughtered my dearest pets. Half-inch mesh hardware cloth would not exclude house mice, but that thought did not bother me at the time, for I simply shrugged off as unimportant the fact that they would cost me something by sharing the pigeon food.

Now I wish I had used a finer mesh, though not for economic reasons. At first I thought that the arrangement was actually a good one, for the house remained unwontedly vermin-free while so much provender lay spread in the shelter nearby. But the lull was misleading. Mice were making nests everywhere out there— under the cement floor, in the tin roof, behind water and grit dispensers, and inside articles stored in an annex to the pigeon shed. There they cut a hammock and a foam-rubber-filled mattress to pieces. When winter came, they simply flooded the house, and that has become a chronic condition. My records show that in 1969 I trapped 55 house mice on the stove, meanwhile catching 19 others among the pigeons. During the first half of 1970 I concentrated on destroying mice at their source and accounted for 109 in the pigeon shed; only 6 got as far as the stove during that period. In July I gratefully gave Kenneth Godfrey, my erstwhile provider of turtles, permission to take over the mouse trapping in order to feed a pet Barn Owl that he was rearing, and he undoubtedly brought the population down to a hard core. But the owl, alas, grew up and flew away, and by October the mice were back as thick as ever. It is obvious that I am stuck with them.

I believe that house mice became an actual detriment to one of my birds, though I can not prove it absolutely. But otherwise I am at a loss to explain how the female Fantail's tail became so badly gnawed. At one time, while I was crossbreeding her with the male Nun, she made a nest on the floor, rather than on one of the shelves I had provided. This happened at a time when I had

not blitzkrieged the mice for quite a while and they were consequently at a populational peak. The Fantail's nest was against a
wall where the mice had established one of their major runways,
so that each mouse slinking by must practically brush against the
bird's spreading tail. Mice are largely nocturnal, and female pigeons customarily incubate their eggs at night, husbands taking a
turn by day. The female's tail became chewed, the Nun's did
not. These circumstances make me believe that the mice simply
nipped off part of a feather in passing, perhaps out of boredom.
But by the time the Fantail finished rearing that pair of squabs,
half her tail was gone.

You might doubt that a pigeon would stay still while under
such attack. I personally think it quite possible, for I regret to
confess that I regard them as highly stupid, despite my love for
them. I repeat that I can not vouch for their responses to mice,
but I did on one occasion see a pair that possessed half-grown
squabs stand quietly by while a black rat snake slithered past
their nest. It turned out that the snake had sensed the presence of
a brood of mice under the water dispenser and was not intent on
squabs—though the parents could not have been aware of that.
The fact is that they simply did not react at all to what must traditionally be a deadly threat to pigeondom.

Snakes were my constant allies in the war on mice. I was
able to identify them as black rat snakes by examining their shed
skins under a lens and noting the longitudinal keel on each scale.
It is impossible to say how many snakes I harbored or how many
mice they ate. However, I have not yet found one in the house,
though if they begin to get smart I suppose that may be the next
novelty.

Eventually a few deer mice succeeded, one by one, in getting past the house-mouse barrier. Only rarely did I catch one on
the stove, for competition with house mice was too keen there.
Rather, they came upstairs, where they could make their own rules.

And strange indeed were those customs! Once, when I decided to put on a pair of sneakers that I had not worn for a
while, I found that I could not get my foot fully into the left one.
The toe was crammed full of seeds. Another time the same thing
happened to one of my bedroom slippers. Both times, I set a Havahart trap and intercepted a white-footed mouse. The combined
harvests must have totalled more than 150 seeds, somewhat

smaller than dried peas, all of the same kind. But they came from a plant, shrub or tree which I have to this day not identified. I canvassed the plantings around the house and all vegetation in nearby woods to discover where the mice had found these *abundant* little nutlike fruits, but could find nothing that might have produced them. In spring I planted the seeds in a row in the vegetable garden, for surely I could make a diagnosis when I had stems and leaves before me. Like a fool I planted them *all;* none came up, and I could no longer locate any seeds in the ground that fall, so I had nothing left to submit to a botanist. I am still wondering what they could be.

Mrs. Fisher called me one Tuesday when I happened to be in Swarthmore and said that thieves had broken into the house. When she had come for her weekly dishwashing and cleaning, she had found the glass of the back door smashed. The thief or thieves had then reached in and turned the lock.

By the time I arrived in Eldora the police had gone over the place thoroughly. They had even brought their fingerprint expert, who pronounced the thief a professional because he had worn gloves. But they were all puzzled by the appearance of things. The thief had gathered loot from my bedroom upstairs and from various rooms downstairs and had piled it on the dining room table. But having thus brought his cache together in readiness for carrying it away, he had suddenly bolted. Heavy footprints in a flower bed showed where he had leapt from the front porch. He had dropped a tire iron on the lawn in his flight—the instrument with which he had broken the glass, and that probably would have doubled as a weapon in case of need. But nothing was missing except for a couple of baubles that he could have slipped into a pocket. What kind of job was *that?*

A theft always makes you a bit jittery for the time being, though you soon get over it. That evening I was filling out a form about damage to the back door for the insurance company. Everything was very quiet until suddenly I jumped about a mile.

> Scramble, scamper, bump! Scamper, bump, scramble!
> Thump, thump, thump, *thump!*
> Scramble, scamper, bump! Scamper, bump, scramble!
> Thump, thump, thump, *thump!*

So *that* was how the geese had saved Rome!

INDEX

This index, prepared by the author, is not a conventional one in that its chief function is to present scientific names of organisms, rather than burdening the text with them. Other topics of discussion, naturalists' names, and so forth were deemed of an incidental nature and are generally not included.

Sphinx. *See* Hawkmoth.
Squirrel, flying. *Glaucomys volans,* 252
Squirrel, gray. *Sciurus carolinensis,* 251
Stinkbug, predatory. Probably of Sub-
family Asopinae in Family Pentato-
midae, 107, 108, 110
Stinkpot. See Turtle, musk
Swallow, Tree. *Iridoprocne bicolor,* 189
Sweet gum. *Liquidambar styraciflua,* 99

Tabanidae. A family of flies including
both horse flies and deer flies among
others, 38
Terrapin, northern diamondback. *Mala-
clemys terrapin terrapin,* 160, 177, 203
Trematode. A parasitic unsegmented
flatworm or fluke, 72
Turtle, Blanding's. *Emydoidea blandingi,*
24
Turtle, diamondback. *See* Terrapin
Turtle, eastern box. *Terrapene carolina
carolina,* 20, 24, 159, 168
Turtle, eastern mud. *Kinosternon subrubrum subrubrum,* 162
Turtle, musk. *Sternothaerus odoratus,*
164
Turtle, painted. *Chrysemys picta,* 24,
161

Turtle, red-eared. *Pseudemys scripta
elegans,* 179
Turtle, snapping. *Chelydra serpentina,*
167, 168

Uranotaenia, 15

Virus, pox (of slate-colored Junco), 72
Vision, in mosquitoes, 27
Vole, boreal redback. *Cleithronomys
gapperi,* 251
Vole, meadow. *Microtus pennsylvanicus,*
251
Vole, pine. *Pitymys pinetorum,* 251

Walnut, black. *Juglans nigra,* 99
Wasp, ichneumon. A wasp of Super-
family Ichneumonoidea that lays eggs
in the bodies of other insects, frequently caterpillars. The wasp's larvae
then develop as internal parasites, 150
Weasel, longtail. *Mustela frenata,* 128
Willet. *Catoptrophorus semipalmatus,*
160
Wyeomyia, 15

Yellowthroat. *Geothlypis trichas,* 212